Tucholsky Wagner Zola Scott Sydow Freud Schlegel
Turgenev Wallace Fonatne
Twain Walther von der Vogelweide Fouqué Friedrich II. von Preußen
Weber Freiligrath Frey
Fechner Weiße Rose von Fallersleben Kant Ernst Frommel
Fichte Richthofen
Hölderlin
Engels Fielding Eichendorff Tacitus Dumas
Fehrs Faber Flaubert
Eliasberg Ebner Eschenbach
Feuerbach Maximilian I. von Habsburg Fock Eliot Zweig
Ewald Vergil
Goethe London
Elisabeth von Österreich
Mendelssohn Balzac Shakespeare Dostojewski Ganghofer
Lichtenberg Rathenau Doyle Gjellerup
Trackl Stevenson Hambruch
Mommsen Tolstoi Lenz Droste-Hülshoff
Thoma von Arnim Hanrieder
Dach Verne Hägele Hauff Humboldt
Karrillon Reuter Rousseau Hagen Hauptmann
Garschin Baudelaire Gautier
Damaschke Defoe Hebbel
Descartes Hegel Kussmaul Herder
Wolfram von Eschenbach Schopenhauer Rilke George
Bronner Darwin Dickens Grimm Jerome Bebel
Campe Melville Aristoteles Federer Proust
Bismarck Vigny Barlach Voltaire Herodot
Gengenbach Heine
Storm Casanova Tersteegen Gilm Grillparzer Georgy
Chamberlain Lessing Langbein Gryphius
Brentano Lafontaine
Strachwitz Claudius Schiller Kralik Iffland Sokrates
Bellamy Schilling
Katharina II. von Rußland Raabe Gibbon Tschechow
Gerstäcker
Löns Hesse Hoffmann Gogol Wilde Gleim Vulpius
Luther Heym Hofmannsthal Klee Hölty Morgenstern Goedicke
Roth Heyse Klopstock Puschkin Homer Kleist
Luxemburg La Roche Horaz Mörike Musil
Machiavelli Kierkegaard Kraft Kraus
Navarra Aurel Musset Lamprecht Kind Moltke
Nestroy Marie de France Kirchhoff Hugo
Nietzsche Nansen Laotse Ipsen Liebknecht
Marx Lassalle Gorki Klett Leibniz Ringelnatz
von Ossietzky May Lawrence Irving
vom Stein
Petalozzi Knigge
Platon Pückler Michelangelo Kock Kafka
Sachs Poe Liebermann
de Sade Praetorius Mistral Zetkin Korolenko

The publishing house tredition has created the series **TREDITION CLASSICS**. It contains classical literature works from over two thousand years. Most of these titles have been out of print and off the bookstore shelves for decades.

The book series is intended to preserve the cultural legacy and to promote the timeless works of classical literature. As a reader of a **TREDITION CLASSICS** book, the reader supports the mission to save many of the amazing works of world literature from oblivion.

The symbol of **TREDITION CLASSICS** is Johannes Gutenberg (1400 – 1468), the inventor of movable type printing.

With the series, tredition intends to make thousands of international literature classics available in printed format again – worldwide.

All books are available at book retailers worldwide in paperback and in hardcover. For more information please visit: www.tredition.com

tredition was established in 2006 by Sandra Latusseck and Soenke Schulz. Based in Hamburg, Germany, tredition offers publishing solutions to authors and publishing houses, combined with worldwide distribution of printed and digital book content. tredition is uniquely positioned to enable authors and publishing houses to create books on their own terms and without conventional manufacturing risks.

For more information please visit: www.tredition.com

Old English Patent Medicines in America

George B. Griffenhagen

Imprint

This book is part of the TREDITION CLASSICS series.

Author: George B. Griffenhagen
Cover design: toepferschumann, Berlin (Germany)

Publisher: tredition GmbH, Hamburg (Germany)
ISBN: 978-3-8495-0507-3

www.tredition.com
www.tredition.de

Copyright:
The content of this book is sourced from the public domain.

The intention of the TREDITION CLASSICS series is to make world literature in the public domain available in printed format. Literary enthusiasts and organizations worldwide have scanned and digitally edited the original texts. tredition has subsequently formatted and redesigned the content into a modern reading layout. Therefore, we cannot guarantee the exact reproduction of the original format of a particular historic edition. Please also note that no modifications have been made to the spelling, therefore it may differ from the orthography used today.

OLD ENGLISH PATENT MEDICINES IN AMERICA

by George B. Griffenhagen and James Harvey Young

r 10, pages 155-183, from

NTRIBUTIONS FROM THE MUSEUM
HISTORY AND TECHNOLOGY

NITED STATES NATIONAL MUSEUM
BULLETIN 218

ITHSONIAN INSTITUTION • WASHINGTON, D.C., 1959

OLD ENGLISH PATENT MEDICINES IN AMERICA

By George B. Griffenhagen and James Harvey Young

Bateman's Pectoral Drops, Godfrey's Cordial, Turlington's Balsam of Life, Hooper's Female Pills, and a half-dozen other similar nostrums originated in England, mostly during the first half of the 18th century. Advertised with extravagant claims, their use soon spread to the American Colonies.

To the busy settler, with little time and small means, these ready-made and comparatively inexpensive "remedies" appealed as a solution to problems of medical and pharmaceutical aid. Their popularity brought forth a host of American imitations and made an impression not soon forgotten or discarded.

The Authors: *George B. Griffenhagen, formerly curator of medical sciences in the Smithsonian Institution's U.S. National Museum, is now Director of Communications for the American Pharmaceutical Association. James Harvey Young is professor of history at Emory University. Some of the material cited in the paper was found by him while he held a fellowship from the Fund for the Advancement of Education, in 1954-55, and grants-in-aid from the Social Science Research Council and Emory University, in 1956-57.*

In 1824 there issued from the press in Philadelphia a 12-page pamphlet bearing the title, *Formulae for the preparation of eight patent medicines, adopted by the Philadelphia College of Pharmacy*. The College was the first professional pharmaceutical organization established in America, having been founded in 1821, and this small publication was its first venture of any general importance. Viewed from the perspective of the mid-20th century, it may seem strange if not shocking that the maiden effort of such a college should be municip-

ing formulas for nostrums. Adding to the novelty is the fact that all eight of these patent medicines, with which the Philadelphians concerned themselves half a century after American independence, were of English origin.

Hooper's Female Pills, Anderson's Scots Pills, Bateman's Pectoral Drops, Godfrey's Cordial, Dalby's Carminative, Turlington's Balsam of Life, Steer's Opodeldoc, British Oil — in this order do the names appear in the Philadelphia pamphlet — all were products of British therapeutic ingenuity. Across the Atlantic Ocean and on American soil these eight and other old English patent medicines, as of the year when the 12-page pamphlet was printed, had both a past and a future.

Origin of English Patent Medicines

When the Philadelphia pharmacists began their study, the eight English patent medicines were from half a century to two centuries old.[1] The most ancient was Anderson's Scots Pills, a product of the 1630's, and the most recent was probably Dalby's Carminative, which appeared upon the scene in the 1780's. Some aspects of the origin and development of these and similar English proprietaries have been treated, but a more thorough search of the sources and a more integrated and interpretive recounting of the story would be a worthy undertaking. Here merely an introduction can be given to the cast of characters prior to their entrances upon the American stage.

The inventor of Anderson's Scots Pills was fittingly enough a Scot named Patrick Anderson, who claimed to be physician to King Charles I. In one of his books, published in 1635, Anderson extolled in Latin the merits of the Grana Angelica, a pill the formula for which he said he had learned in Venice. Before he died, Anderson imparted the secret to his daughter Katherine, and in 1686 she in turn conveyed the secret to an Edinburgh physician named Thomas Weir. The next year Weir persuaded James II to grant him letters patent for the pills. Whether he did this to protect himself against competition that already had begun, or whether the patenting gave a cue to those always ready to cut themselves in on a good thing, cannot be said for sure. The last years of the 17th century, at any rate, saw the commencement of a spirited rivalry among various makers of Anderson's Scots Pills that was long to continue. One of them was Mrs. Isabella Inglish, an enterprising woman who sealed her pill boxes in black wax bearing a lion rampant, three mallets argent, and the bust of Dr. Anderson. Another was a man named Gray who sealed his boxes in red wax with his coat of arms and a motto strangely chosen for a medicine, "Remember you must die."

FORMULÆ

FOR THE

PREPARATION OF EIGHT

PATENT MEDICINES,

ADOPTED BY THE

PHILADELPHIA COLLEGE

OF

PHARMACY

MAY 4th, 1824.

SOLOMON W. CONRAD, PRINTER,

No. 32, Church Alley.

Figure 1.—The Philadelphia College of Pharmacy in 1824 set forth in this pamphlet formulas for eight old English patent medicines. (*Courtesy, Philadelphia College of Pharmacy and Science, Philadelphia, Pennsylvania.*)

Competition already had begun when Godfrey's Cordial appeared in the record in a London newspaper advertisement during December 1721. John Fisher of Hertfordshire, "Physician and Chymist," claimed to have gotten the true formula from its originator, the late Dr. Thomas Godfrey of the same county. But there is an alternate explanation. Perhaps the Cordial had its origin in the apothecary shop established about 1660 by Ambroise (Hanckowitz) Godfrey in Southampton Street, London.[2] According to a handbill issued during the late 17th century, Ambroise Godfrey prepared "Good Cordials as Royal English Drops."

Elixir Salutis:

THE
CHOISE DRINK OF HEALTH,

OR,

𝔥𝔢𝔞𝔩𝔱𝔥-𝔅𝔯𝔦𝔫𝔤𝔦𝔫𝔤 𝔇𝔯𝔦𝔫𝔨.

BEING

A Famous Cordial Drink, found out by the Providence of the Almighty, and Experienced a Most Excellent Preservative of Man-kind,

A SECRET

Far beyond any Medicament yet known, And is found so agreeable to Nature, That it Effects all its Operations, as Nature would Have it, and as a Vertual Expedient proposed by her, for reducing all her Extreams unto an equal Temper; the same being fitted unto all *Ages*, *Sexes*, *Complexions* and *Constitutions*, and highly fortifying Nature against any Noxious humour, invading or offending the *Noble-Parts:*

Never Published by any but by Me
ANTHONY DAFFY, Student in *Physick*.

LONDON.
Printed with Allowance for the Author by T. Milbourn, 1673.

Figure 2.—Anthony Daffy Extolled the Virtues of His Elixir Salutis in this pamphlet, published in London in 1673. (*Courtesy, British Museum.*)

With respect to his rivals, the 18th-century Hertfordshire vendor of the Cordial warned in the *Weekly Journal* (London), December 23, 1721: "I do advise all Persons, for their own Safety, not to meddle with the said Cordial prepared by illiterate and ignorant Persons, as Bakers, Malsters, [sic] and Goldsmiths, that shall pretend to make it, it being beyond their reach; so that by their Covetousness and Pretensions, many Men, Women, and especially Infants, may fall as Victims, whose Slain may exceed Herod's Cruelty...."

In 1726 King George I granted a patent for the making and selling of Dr. Bateman's Pectoral Drops. The patent was given not to a doctor, but to a business man named Benjamin Okell. In the words of the patent,[3] Okell is lauded for having "found out and brought to Perfection, a new Chymicall Preparacion and Medicine..., working chiefly by Moderate Sweat and Urine, exceeding all other Medicines yet found out for the Rheumatism, which is highly useful under the Afflictions of the Stone, Gravell, Pains, Agues, and Hysterias...." What the chemicals constituting his remedy were, the patentee did not vouchsafe to reveal.

The practice of patenting had begun in royal prerogative. Long accustomed to granting monopoly privileges for the development of new industries, the discovery of new lands, and the enrichment of court favorites, various monarchs in 17th-century Europe had given letters patent to proprietors of medical remedies which had gained popular acclaim. In France and the German States, this practice continued well through the 18th century. In England, where representative government had progressed at the expense of the personal prerogative of the sovereign, Parliament passed a law in 1624 aimed at curbing arbitrary actions like those of James I and Charles I. The statute declared all monopolies void except those extended to the first inventor of a new process of manufacture. To such pioneers the king could grant his letters patent bestowing monopoly privileges for a period of 14 years. That the machinery set up by this law did not completely curb the independence of English sovereigns in the medical realm is indicated by the favor extended Dr. Weir, who successfully sought from James II a privileged position for Anderson's Scots Pills. This kingly grant is not included in the regular list, and the Glorious Revolution of 1688 brought an end to such an exercise of royal power without consent of Parliament. A list of patents in the medical field later published by the Commissioners of Patents[4] includes only six issued during the 17th century, four for baths and devices, one for an improved method of preparing alum, and one for making epsom salts. The first patent for a compound medicine was granted in 1711, and only two other proprietors preceded Benjamin Okell in seeking this particular legal form of protection and promotion.

As early as 1721, Bateman's Pectoral Drops were being regularly advertised in the *London Mercury*. The advertisements announced: "Dr. Bateman's Pectoral drops published at the Request of several Persons of Distinction from both Universities...." The Drops, priced at "1 s. a Bottle," were "Sold Wholesale and Retail at the Printinghouse and Picture Warehouse in Bow Churchyard," and likewise "in most Cities and celebrated Towns in Great Britain." "Each Bottle Seal'd with the Boar's Head." So stated the advertisement, which itself contained a crude cut of this Boar's Head seal.[5] Elsewhere in this issue of the *Mercury*, we learn that John Cluer, printer, was the proprietor of the Bow Churchyard Warehouse. This same John Cluer, along with William Dicey and Robert Raikes, were named in the 1726 patent as "the Persons concerned with the said Inventor," Benjamin Okell, who, with him, should "enjoy the sole Benefit of the said Medicine." It was this partnership which was to find the field of nostrum promotion especially congenial and which was to play an important transatlantic role. Soon after securing their patent, the proprietors undertook to inform their countrymen about the remedy by issuing *A short treatise of the virtues of Dr. Bateman's Pectoral Drops*.[6]

It was the 18th century, and the essay was in fashion. The proprietors prepared a didactic introduction to their treatise, phrased in long and flowery sentences, in which modesty was not the governing tone. The arguments ran like this: that the "Universal Good of Mankind" should be the aim of "every private member"; that nothing is so conducive to this general welfare as "health"; that no hazards to health are more direful than diseases such as "the Gout; the

Rheumatism; the Stone; the Jaundice," etc., etc.; that countless men and women have succumbed to such afflictions either because they received no treatment or suffered wrong treatment at "the Hands of the Learned"; that no medicine is so sure a cure as that inexpensive remedy discovered as a result of great "Piety, Learning and Industry" by one "inspir'd with the Love of his Country, and the Good of Mankind," to wit. "Dr. bateman's Pectoral Drops."

Then followed seven chapters treating the multitude of illnesses for which the Drops were a specific. Finally, the pamphlet cited "some few, out of the many thousands of Certificates of Cures effected by these drops...." Even so early was the testimonial deemed a powerful persuader.

No more could Okell, Cluer, Dicey, and Raikes escape competition than could the proprietors of other successful nostrums. In 1755 they went to court and won a suit for the infringement of their patent, but the damages amounted to only a shilling. Even after the patent expired, the tide of publicity flowed on.[7]

Competition was also lively in the 1740's among some half a dozen proprietors marketing a form of crude petroleum under the name of British Oil. Early in the decade Michael and Thomas Betton were granted a patent for "An Oyl extracted from a Flinty Rock for the Cure of Rheumatick and Scorbutick and other Cases." The source of the oil, according to their specifications, was rock lying just above the coal in mines, and this rock was pulverized and heated in a furnace to extract all the precious healing oil.[8] This Betton patent aroused one of their rivals, Edmund Darby & Co. of Coalbrook-Dale in Shropshire. Darby asserted that it was presumptuous of the Bettons to call their British oyl a new invention.[9] For over a century Darby and his predecessors had been marketing this self-same product, and it had proved to be "the one and only unrivall'd and most efficacious Remedy ever yet discovered, against the whole force of Diseases and Accidents that await Mankind...." For the Bettons to appropriate the process and patent it—and even to claim in their advertising cures which really had been wrought by the Darby product—was scandalous. Worse than that, said Darby, it was illegal, for in 1693 William III had granted a patent to "Martin Eele and two others at his Nomination for making the same Sort of

Oyl from the same Sort of Materials." Evidence to substantiate his belief in the Betton perfidy was presented by Darby to George II, who had the matter duly investigated.[10] Being persuaded that Darby was right, the king and his councillors, in 1745, vacated the Betton patent. This victory seems not to have boomed the Darby interests, and this defeat seems not to have ruined the Bettons. During the succeeding century, the Betton patent was published and republished in advertising, just as if it had never fallen afoul the law. From their battles with the Oil from Coalbrook-Dale and other British Oils marketed by other proprietors, the Bettons emerged triumphant. In the years to come, patent or no, the Bettons British Oil was to dominate the field.

The year after the Bettons had secured their patent, another was granted to John Hooper of Reading for the manufacture of "Female Pills" bearing his name.[11] Hooper was an apothecary, a man-midwife, and a shrewd fellow. This was the period in which the British Government was increasing its efforts to require the patentee to furnish precise specifications with his application.[12] When Hooper was called upon to tell what was in his pills and how they were made, he replied by asserting that they were composed "Of the best purging stomatick and anti-hysterick ingredients," which were formed into pills the size of a small pea. This satisfied the royal agents and Hooper went on about his business. In an advertisement of the same year, he was able to cite as a witness to his patent the name of the Archbishop of Canterbury.[13]

Much less taciturn than Hooper about the composition of his nostrum was Robert Turlington, who secured a patent in 1744 for "A specifick balsam, called the balsam of life."[14] The Balsam contained no less than 27 ingredients, and in his patent specifications Turlington asserted that it would cure kidney and bladder stones, cholic, and inward weakness. He shortly issued a 46-page pamphlet in which he greatly expanded the list.[15] In this appeal to 18th-century sensibilities, Turlington asserted that the "Author of Nature" has provided "a Remedy for every Malady." To find them, "Men of Learning and Genius" have "ransack'd" the "Animal, Mineral and Vegetable World." His own search had led Turlington to the Balsam, "a perfect Friend to Nature, which it strengthens and corroborates when weak and declining, vivifies and enlivens the Spirits, mixes with the Juices and Fluids of the Body and gently infuses its kindly Influence into those Parts that are most in Disorder."

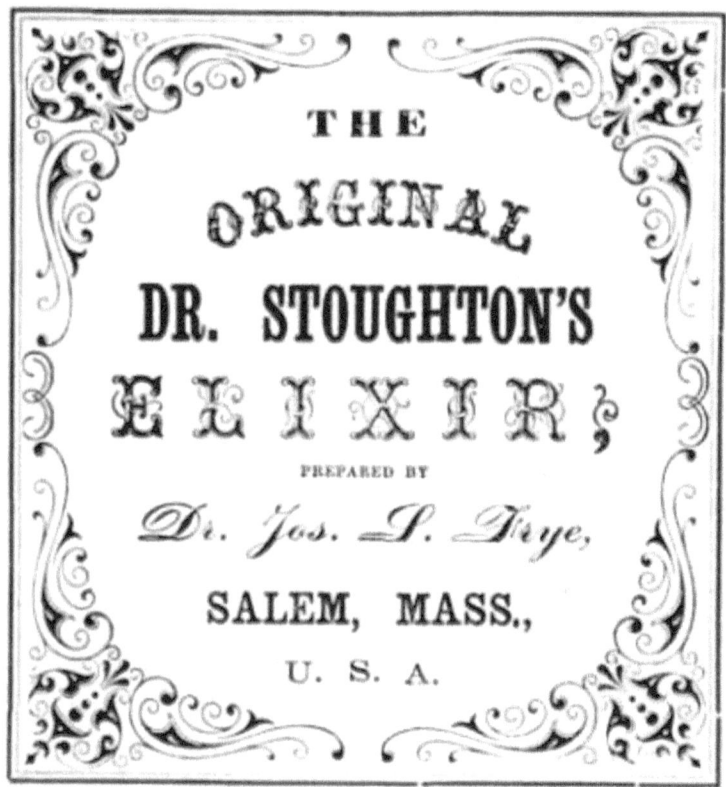

Figure 3.—Label for Stoughton's Elixir as manufactured by Dr. Jos. Frye of Salem, Massachusetts. (*Courtesy, Essex Institute, Salem, Massachusetts.*)

Testimonials from those who had felt the kindly influence took up most of the space in Turlington's pamphlet. In these grateful acknowledgments to the potency of the patent medicine, the list of illnesses cured stretched far beyond the handful named in the patent specifications. Just as for Bateman's Pectoral Drops and the Darby brand of British Oil, workers of many occupations solemnly swore that they had received benefit. Most of them were humble people—a porter, a carpenter, the wife of a gardener, a blanket-weaver, a gunner's mate, a butcher, a hostler, a bodice-maker. Some bore a status of greater distinction: there were a "Mathematical Instrument-Maker" and the doorkeeper of the East India Company. All were jubilant at their restored good health.

The Balsam's well-nigh sovereign power could not protect it from one ailment of the times, competition. Various preparations of similar composition, like Friar's Balsam, already were on the market, but before long even the Turlington name was trespassed upon, and the inventor's niece was forced to advertise that she alone had the true formula and that any person who took a dose of the spurious imitations being offered did so at great hazard to his life.

A quarter of a century after the patenting of the Balsam, there appeared for sale to British ailing a remedy called Dr. Steer's Celebrated Opodeldoc. Dr. Steer is a shadowy rider of a vigorous steed, for although the doctor has left but a faint personal impact upon the historical record, Opodeldoc has pranced through medical history since the time of Paracelsus. This 16th-century continental chemist-physician, who introduced many mineral remedies into the materia medica, had coined the word "opodeldoc" to apply to various medical plasters. In the two ensuing centuries the meaning had changed, and the *Pharmacopoeia Edinburgensis* of 1722 employed the term to designate soap liniment. It is presumed that Dr. Steer appropriated the Edinburgh formula, added ammonia, and marketed his proprietary version. In 1780, a London paper carried an advertisement listing the difficulties for which the Opodeldoc was a "speedy and certain cure." These included bruises, sprains, burns, cuts, chillblains, and headaches. Furthermore, the remedy had been "found of infinite Use in hot Climates for the Bite of venomous Insects."[16] Dr. Steer seems not to have secured a patent for his slightly modified version of an official preparation. He died in 1781, but Opodeldoc, indeed Steer's Opodeldoc, went marching on.[17]

About the same time that Dr. Steer began advertising, newspaper promotion was launched in behalf of another remedy, called Dalby's Carminative. The inventor, J. Dalby, was a London apothecary, and his unpatented concoction was designed to cure "Disorders of the Bowels." One early advertisement[18] added details: "This Medicine, which is founded on just Medical Principles, has been long established as a most safe and effectual Remedy, generally affording immediate Relief in the Wind, Cholocks [sic], Convulsions, Purgings, and all those fatal Disorders in the Bowels of Infants, which carry off so great a number under the age of 2 years. It is also equally efficacious in gouty Pains in the Intestines, in Fluxes, and in the cholicky Complaints of grown Persons, so usual at this Season of the Year." Dalby, like Steer, failed long to survive the appearance of his medicine on the market.

Such were the origins of the eight remedies which the Philadelphia pharmacists were to take account of in 1824. Besides these eight, two other patent medicines, both elixirs, were destined for roles of such special interest that a brief look at their English background is warranted.

One of them, Daffy's Elixir, was the invention of a clergyman, Rev. Thomas Daffy soon after 1650. Daffy had his troubles during that troubled century, losing a pastorate because he offended a powerful Countess. When the rector first sought to minister unto men's bodies as well as to their souls is not known. According to a pamphlet issued in 1673, after the Rev. Daffy had passed from the scene, the formula had been "found out by the Providence of the Almighty." By this time a London kinsman of the inventor, named Anthony Daffy, was vending the remedy. The full name of the medicine, according to the pamphlet's title, was "Elixir Salutis: The Choice Drink of Health, or Health-Bringing Drink," and among the ailments for which it was effective were gout, the stone, colic, "ptissick," scurvy, dropsy, rickets, consumption, and "languishing and melancholly."

The Elixir Salutis proved immensely popular. It was too much to expect that Anthony should hold the field uncontested; in the 1673 pamphlet one false fabricator was called by name, and in 1680 Anthony advertised to warn against "diverse Persons" who were not only counterfeiting the medicine but spreading the malicious rumor that Anthony was dead. Early in the new century, Catherine, the daughter of the original Rev. Daffy, insisted that she as well as her cousin Anthony had received the valuable formula. But it was Anthony's line that was to prove the more persistent. In 1743, one Susannah Daffy advertised the "Original and Famous Elixir," asserting that she had a brother Anthony who also knew the secret.[19] This Anthony died in 1750 and willed the formula to his niece. But there were others outside the family who long had been making and selling the medicine. For example, the Bow Churchyard Warehouse advertised Daffy's Elixir in the *London Mercury* during 1721. Without hiding the fact that others were also compounding this "safe and pleasant Cordial ... well-known throughout England, where it has been in great Use these 50 Years," the advertisement concluded: "Those who make tryalof That sold at this [Bow Churchyard] Warehouse will never buy anywhere else."[20]

Although once lauded by a physician to King Charles II, Daffy's Elixir was never patented. The Elixir invented by Richard Stoughton was, in 1712, the second compound medicine to be granted a patent in England.[21] Stoughton was an apothecary who had a shop at the Sign of the Unicorn in Southwark, Surrey. It was evidently competition, the constant bane of the medicine proprietor's life, that drove him to seek governmental protection. In his specifications he asserted that he had been making his medical mixture for over twenty years. Stoughton was less precise about his formula; indeed, he gave none, but was generous in indicating the remedy's name: "Stoughton's Elixir Magnum Stomachii, or the Great Cordial Elixir, otherwise called the Stomatick Tincture or Bitter Drops." In a handbill, the apothecary did tip his hand to the extent of asserting that his Elixir contained 22 ingredients, but added that nobody but himself knew what they were. The dosage was generous, 50 to 60 drops "in a glass of Spring water, Beer, Ale, Mum, Canary, White wine, with or without sugar, and a dram of brandy as often as you please." This, it was said, would cure any stomach ailment whatever.[22]

The inventor died in 1726, and his passing precipitated a perfect fury of competitive advertising. As in the case of Daffy's, there was a family feud. A son of Stoughton and the widow of another son argued vituperously in print, each claiming sole possession of Richard's complicated secret, and each terming the other a scoundrel. The daughter-in-law accused the son of financial chicanery, and the son condemned the daughter-in-law for having run through two husbands and for desperately wanting a third. In the midst of this running battle, a third party entered the lists as maker of the Elixir. She was no Stoughton—though a widow—and her quaint claim for the public's consideration lay in this, that her late husband had infringed Stoughton's patent until restrained by the Lord Chancellor.

These ten medicines—Stoughton's and Daffy's Elixirs and the eight which the Philadelphia pharmacists were later to select—were by no means the only packaged remedies available to the 18th-century Englishman who resorted to self-dosage for his ills. Between 1711, when the first patent was granted for a compound medicine, and 1776, some 75 items were patented in the medical field.[23] And, along with Godfrey's Cordial and Daffy's Elixir, there were scores of other remedies for which no patents had been given. A list of nostrums published in *The Gentleman's Magazine* in 1748 totaled 202, and it was admittedly incomplete.[24] The proprietor with a patent might do his utmost to keep this badge of governmental sanction before the public, but the distinction was not great enough in such a crowded field to make things clear. The casual buyer could not keep track of which electuary had been granted a patent and which lozenge had not. They were all bottles and boxes upon the shelf. In use they served the same purpose. One term arose in common speech to apply to both, and it was "patent medicine."

English Patent Medicines Come to America

When the first English packaged medicine, patented or unpatented, came to the New World, cannot be told. Some 17th-century prospective colonist, setting forth to face the hazards of life in Jamestown or Baltimore or Boston, must have packed a box of Anderson's Scots Pills or a bottle of Daffy's Elixir to bring along, but no record to substantiate such an incident has been encountered. It would seem that the use of English packaged remedies in America was most infrequent before 1700. Samuel Lee, answering questions posed from England in 1690 about the status of medicine and pharmacy in Massachusetts, mentions no patent medicines.[25] Neither does the 1698 account book of the Salem apothecary, Bartholomew Brown.[26]

Figure 4.—Patrick Anderson, M.D., from a box of Anderson's Scots Pills. From Wootton's *Chronicles of pharmacy*, London, 1910. (*Smithsonian photo 44286-C.*)

In the *Boston News-Letter* for October 4, 1708, Nicholas Boone, at the Sign of the Bible, near the corner, of School-House-Lane, advertised for sale: "Daffy's Elixir Salutis, very good, at four shillings and sixpence *per* half pint Bottle." This may well be the first printed reference in America to an English patent medicine, and it certainly is the first newspaper advertisement for a nostrum. Preceding the *News-Letter* in colonial America, there had been only one paper, the *Publick Occurrences Both Foreign and Domestic*.[27] This journal had lasted but a single issue. Then its printer had returned to England, where he took up the career of a patent medicine promoter, vending "the only Angelical Pills against all Vapours, Hysterick Fits." The *News-Letter* had begun with the issue of April 27, 1704, about 4 years before Boone's advertisement for Daffy's remedy made its appearance, but during that time, only one advertisement for anything at all in the medical field had appeared, and that was for a home-remedy book, *The English physician*, by Nicholas Culpeper, Doctor of Physick.[28] This volume was also for sale at Boone's shop.

Patent-medicine advertising in the *News-Letter* prior to 1750 was infrequent. Apothecary Zabdiel Boylston, who a decade later was to earn a role of esteem in medical history by introducing the inoculation for smallpox, announced in 1711 that he would sell "the true Lockyers Pills."[29] This was an unpatented remedy first concocted half a century earlier by a "licensed physitian" in London. The next year Boylston repeated this appeal,[30] and in the same advertisement listed other wares of the same type. He had two varieties, Golden and Plain, of the Spirit of Scurvy-Grass; he had "The Bitter Stomach Drops," worm potions for children; and a wonderful multipurpose nostrum, "the Royal Honey Water, an Excellent Perfume, good against Deafness, and to Make Hair grow...." The antecedents of this regal liquid are unknown. Boylston also announced for sale "The Best [Daffy's] Elixir Salutis in Bottles, or by the Ounce." This is a provocative listing. It may mean merely that the apothecary would break a bottle to sell a dose of the Elixir, which was often the custom. But it also may suggest that Boylston was making the Elixir himself, or was having it prepared by a journeyman. This latter interpretation would place Boylston well at the head of a long parade of American imitators of the old English patent medicines.

Other such shipments of the packaged English remedies may have come to New England on the latest ships from London during the next several decades, but they got scant play in the advertising columns of the small 4-page *Boston News-Letter*. Another reference to "Doctor Anthony Daffey's Original Elixer Salutis" occurs in 1720.[31] Ten years later, Stoughton's Drops were announced for sale "by Public Vendue," along with feather beds, looking glasses, and leather breeches.[32] Nearly a decade more was to pass before Bateman's Pectoral Drops showed up in the midst of another general list, including cheese, and shoes, and stays.[33] Not until 1748 did an advertisement appear in which several of the old English nostrums rubbed shoulders with each other.[34] Then Silvester Gardiner, at the Sign of the Unicorn and Mortar, asserted that "by appointment of the Patentee" he was enabled to sell "Genuine British Oyl, *Bateman's* Pectoral Drops, and *Hooper's* Female Pills, and the True *Lockyer's* Pills."

Although nearly a century old, Anderson's Scots Pills were not cited for sale in the pages of the *Boston News-Letter* until August 23, 1750, two months after the much more recent Turlington's Balsam of Life first put in its appearance.[35] During the same year, the British confusion over British Oil was reflected in America. Boden's and Darby's variety preceded the Betton brand into the *News-Letter* pages by a fortnight.[36] It was the latter, however, which was to win the day in Boston, for almost all subsequent advertising specified the Betton Oil. Godfrey's Cordial was first mentioned in 1761.[37] Thus, of the ten old English patent medicines which are the focus of the present study, eight had been advertised in the *Boston News-Letter*. The other two, Steer's Opodeldoc and Dalby's Carminative, did not reach the market before this colonial journal fell prey to the heightening tensions of early 1776.

By the 1750's, the names of several old English nostrums were appearing fairly frequently in the advertising of colonial apothecaries, not only in Boston but in other colonial towns. In Williamsburg, for example, a steady increase occurs in the number of references and the length of the lists of the English patent medicines advertised in the *Virginia Gazette* from their first mention into the early 1760's.[38] This journal—which later had competing issues by different editors—was launched in 1736, and the next year George Gilmer advised customers that, in addition to "all manner of Chymical and Galenical Medicines," he could furnish, at his old shop near the Governor's, "Bateman's Drops, Squires Elixir, Anderson's Pills."[39] The other remedies appeared in due time, Stoughton's and Daffy's Elixirs in 1745, Turlington's Balsam in 1746, Godfrey's Cordial in 1751, Hooper's Pills in 1752, and Betton's British Oil in 1770.

A spot check of newspapers in Philadelphia and New York reveals a pattern quite similar. Residents of the middle colonies, like those to the north and the south, could buy the basic English brands, and it was during the 1750's that the notices of freshly-arrived supplies ceased to be rare in advertising columns and became a frequent occurrence. Thomas Preston, for example, announced to residents of Philadelphia in 1768 that he had just received a supply of Anderson's, Hooper's, Bateman's, Betton's, Daffy's, Stoughton's, Turlington's, and Godfrey's remedies.[40] Not only were these medicines for sale at apothecary shops, but they were sold by postmasters, goldsmiths, grocers, hair dressers, tailors, printers, booksellers, cork cutters, the post-rider between Philadelphia and Williamsburg, and by many colonial American physicians.

It is a matter for comment that American newspaper advertising of the English packaged medicines was singularly drab. In the mother country, the proprietors or their heirs were faced with vigorous competition. It behooved them to sharpen up their adjectives and reach for their vitriol. In America the apothecary or merchant had no proprietary interest in any of the different brands of the imported medicines which were sold. Moreover, there was probably no great surplus of supply over demand in America as in Britain, so the task of selling the stock on hand was less difficult and required less vigorous promotion. Also, advertising space in the few American weeklies was more at a premium than in the more frequent and numerous English journals. With rare exceptions, therefore, the old English patent medicines were merely mentioned by name in American advertising. Seldom did one receive the individual attention accorded by Samuel Emlen to Godfrey's Cordial in Benjamin Franklin's *Pennsylvania Gazette* for June 26, 1732. The ad ran like this:

"Dr. Godfrey's General Cordial. So universally approved of for the Cholick, and all Manners cf Pains in the Bowels, Fluxes, Fevers, Small-Pox, Measles, Rheumatism, Coughs, Colds, and Restlessness in Men, Women, and Children; and particularly for several Ailments incident to Child-bearing Women, and Relief of young Children in breeding their Teeth."

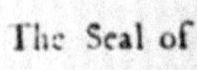

Figure 5.—Pamphlet, Dated 1731, on Behalf of Bateman's Pectoral Drops. It was published by John Peter Zenger in New York. Original preserved in the New York Academy of Medicine Library. (*Smithsonian photo 44286-D.*)

Emlen's venturesomeness may have lain in the fact that he was not only a retailer, but also an agent for the British manufacturer, for he cited the names of those who sold Godfrey's Cordial in nearby towns. Even at that, this appeal, consisting merely of a list of illnesses, lacked the cleverness of contemporary English nostrum advertising. In the whole span of the *Boston News-Letter*, beginning in 1704, it was not until 1763 that a bookstore pulled out the stops with half a column of lively prose in behalf of Dr. Hill's four unpatented nostrums.[41] It seems a safe assumption that not only the medicines but the verbiage were imported from London, where Dr. Hill had been at work endeavoring to restore a Greek secret which "converts a Glass of Water into the Nature and Quality of Asses Milk, with the Balsamick Addition...."

The infrequency of extended fanciful promotion in behalf of the old English nostrums in American newspaper advertising may have been compensated for to some degree in broadside and pamphlet. A critic of the medical scene in New York in the early 1750's asserted that physicians used patent medicines which they learned about from "London quack bills." This doctor complained, these were often their only reading matter.[42] Such a judgment may be too severe. Certainly it is difficult to validate today. Such pamphlets and broadsides do appear in American archival collections. The Historical Society of Pennsylvania contains a 2-page Turlington broadside,[43] while the Folger Shakespeare Library in Washington has an earlier 46-page Turlington pamphlet with testimonials reaching out toward America.[44] One such certificate came from "a sailor before the mast, on board the ship Britannia in the New York trade," and another cited a woman living in Philadelphia who gave thanks for the cure of her dropsy.

A broadside in the Warshaw Collection touting Bateman's Drops noted that "extraordinary demands have been made for Maryland, New-York, Jamaica, etc. where their virtues have been truely experienced with the greatest satisfaction."[45] That such promotional items are extremely rare does not mean they were not abundant in the mid-18th century, for this type of printed matter, then as now, was likely to be looked at and thrown away. A certain amount of nostrum literature was undoubtedly imported from Britain. For example, in 1753 apothecary James Carter of Williamsburg ordered from England "3 Quire Stoughton's Directions" along with "½ Groce Stoughton Vials."[46] These broadsides or circulars served a twofold purpose. Not only did they promote the medicine, but they actually served as the labels for the bottles. Early packages of these patent medicines which have been discovered indicate that paper labels were seldom applied to the glass bottles; instead, the bottle was tightly wrapped and sealed in one of these broadsides.

American imprints seeking to promote the English patent medicines were certainly rare. The most significant example may be found in the Library of the New York Academy of Medicine.[47] In 1731 James Wallace, a New York merchant, became American agent for the sale of Dr. Bateman's Pectoral Drops. To help him with his new venture, Wallace took a copy of the London promotional pamphlet to a New York printer to be reproduced. The printer was John Peter Zenger, not yet an editor and three years away from the events which were to link his name inextricably with the concept of the freedom of the press. This 1731 pamphlet may well have been the earliest work on any medical theme to be printed in New York.[48]

Now and then a physician might frown on his fellows for reading such literature and prescribing such remedies, but he was in a minority. Colonial doctors, by and large, had no qualms about employing the packaged medicines. It was a doctor who first advertised Anderson's Pills and Bateman's Drops in Williamsburg;[49] it was another, migrating from England to the Virginia frontier, who founded a town and dosed those who came to dwell therein with Bateman's Drops, Turlington's Balsam, and other patent medicines.[50]

Complex Formulas and Distinctive Packages

Indeed, the status of medical knowledge, medical need, and medical ethics in the 18th century permitted patent medicines to fit quite comfortably into the environment. As to what actually caused diseases, man knew little more than had the ancient Greeks. There were many theories, however, and the speculations of the learned often sound as quaint in retrospect as do the cocky assertions of the quack bills. Pamphlet warfare among physicians about their conflicting theories achieved an acrimony not surpassed by the competing advertisers of Stoughton's Elixir. The aristocratic practitioners of England, the London College of Physicians, refused to expand their ranks even at a time when there were in the city more than 1,300 serious cases of illness a day to every member of the College. The masses had to look elsewhere, and turned to apothecaries, surgeons, quacks, and self-treatment.[51] The lines were drawn even less sharply in colonial America, and there was no group to resemble the London College in prestige and authority. Medical laissez-faire prevailed. "Practitioners are laureated gratis with a title feather of Doctor," wrote a New Englander in 1690. "Potecaries, surgeons & midwifes are dignified acc[ording] to successe."[52] Such an atmosphere gave free rein to self-dosage, either with an herbal mixture found in the pages of a home-remedy book or with Daffy's Elixir.

In the 18th century, drugs were still prescribed that dated back to the dawn of medicine. There were Theriac or Mithridatum, Hiera Picra (or Holy Bitters), and Terra Sigillata. Newer botanicals from the Orient and the New World, as well as the "chymicals" reputedly introduced by Paracelsus, found their way into these ancient formulas. Since the precise action of individual drugs in relation to given ailments was but hazily known, there was a tendency to blanket assorted possibilities by mixing numerous ingredients into the same formula. The formularies of the Middle Ages encouraged this so-called "polypharmacy." For example the *Antidotarium Nicolai*, written about A.D. 1100 at Salerno, described 38 ingredients in Confectio Adrianum, 35 ingredients in Confectio Atanasia, and 48 ingredients in Confectio Esdra. Theriac or Mithridatum grew in complexity until by the 16th century it had some 60 different ingredients.

It was in this tradition of complex mixtures that most of the patent medicines may be placed. Richard Stoughton claimed 22 ingredients for his Elixir, and Robert Turlington, in his patent specification, named 27. Although other proprietors had shorter lists or were silent on the number of ingredients, a major part of their secrecy really lay in having complicated formulas. Even though rivals might detect the major active ingredients, the original proprietor could claim that only he knew all the elements in their proper proportions and the secret of their blending.

Not only in complexity did the patent medicines resemble regular pharmaceutical compounds of the 18th century. In the nature of their composition they were blood brothers of preparations in the various pharmacopoeias and formularies. Indeed, there was much borrowing in both directions. An official formula of one year might blossom out the next in a fancy bottle bearing a proprietor's name. At the same time, the essential recipe of a patent medicine, deprived of its original cognomen and given a Latin name indicative of its composition or therapeutic nature, might suddenly appear in one of the official volumes.

For example, the formula for Daffy's Elixir was adopted by the *Pharmacopoeia Londinensis* in 1721 under the title of "Elixir Salutis" and later by the *Pharmacopoeia Edinburghensis* as "Tinctura sennae composita" (Compound Senna Tincture). Similarly the essential formula for Stoughton's Elixir was adopted by the *Pharmacopoeia Edinburghensis* as early as 1762 under the name of "Elixir Stomachium," and later as "Compound Tincture of Gentian" (as in the *Pharmacopoeia of the Massachusetts Medical Society* of 1808). Only two years after Turlington obtained his "Balsam of Life" patent, the *Pharmacopoeia Londinensis* introduced a recipe under the title of "Balsamum Traumaticum" which eventually became Compound Tincture of Benzoin, with the synonym Turlington's Balsam. On the other hand, none of these early English patent medicines, including Stoughton's Elixir and Turlington's Balsam, offered anything new, except possibly new combinations or new proportions of ingredients already widely employed in medicine. Formulas similar in composition to those patented or marketed as "new inventions" can in every case be found in such 17th-century pharmacopoeias as William Salmon's *Pharmacopoeia Londinensis*.

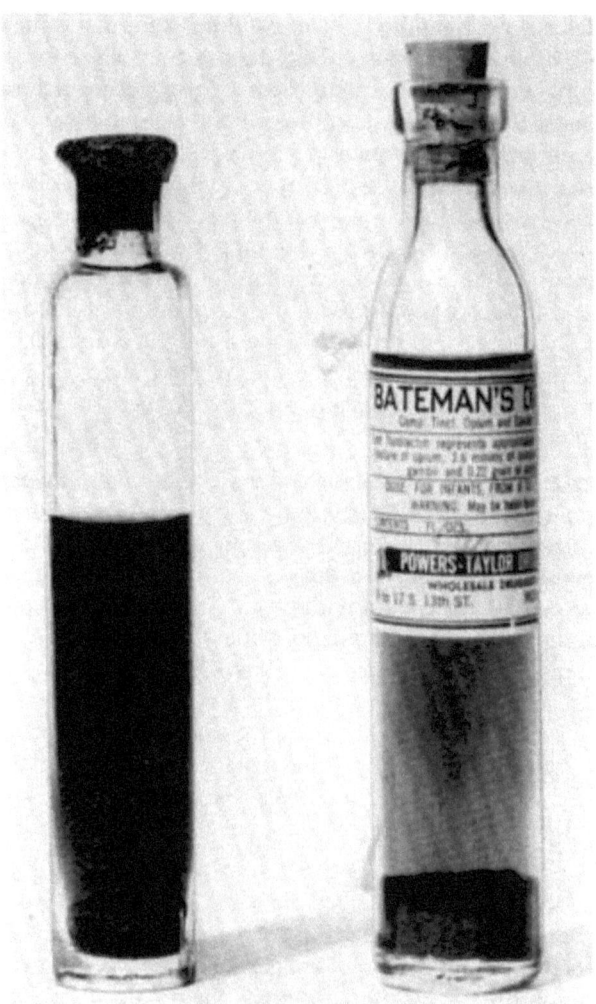

Figure 6.—Bottles of Bateman's Pectoral Drops, 19th century (left) and early 20th century (right), from the Samuel Aker, David and George Kass collection, Albany, New York. (*Smithsonian photo 44287-A.*)

Whatever similarities existed between the canons of regular pharmacy and the composition of patent medicines, there was a decided difference in the methods of marketing. Although patent medicines were often prescription items, they did not have to be. The way they looked on a shelf made them so easily recognizable that even the most loutish illiterate could tell one from another. As the nostrum proprietor did so much to pioneer in advertising psychology, so he also blazed a trail with respect to distinctive packaging. The popularity of the old English remedies, year in and year out, owed much to the fact that though the ingredients inside might vary (unbeknownst to the customer), the shape of the bottle did not. This was the reason proprietors raised such a hue and cry about counterfeiters. The secret of a formula might, if only to a degree, be retained, but simulation of bottle design and printed wrapper was easily accomplished, and to the average customer these externals were the medicine.

This fundamental fact was to be recognized by the committee of Philadelphia pharmacists in 1824. "We are aware" the committeemen reported, "that long custom has so strongly associated the idea of the genuineness of the Patent medicines, with particular shapes of the vials that contain them, and with certain printed labels, as to render an alteration in them an affair of difficulty. Many who use these preparations would not purchase British Oil that was put up in a conical vial, nor Turlington's Balsam in a cylindrical one. The stamp of the excise, the king's royal patent, the seal and coat of arms which are to prevent counterfeits, the solemn caution against quacks and imposters, and the certified lists of incredible cures, [all these were printed on the bottle wrappers] have not even now lost their influence." Nor were they for years to come.

Thus after 1754 the Turlington Balsam bottle was pear-shaped, with sloping shoulders, and molded into the glass in crude raised capitals were the proprietor's name and his claim of the kings royal patent.[53] Turlington during his life had made one modification. He explained it in a broadside, saying that "to prevent the Villainy of some Persons who buying up my empty Bottles, have basely and wickedly put therein a vile spurious Counterfeit-Sort," he had changed the bottle shape. The date molded into the glass on his supply of new genuine bottles was January 26, 1754.[54] This was, perhaps, a very fine point of difference from the perspective of the average customer, and in any case the bottle was hidden under its paper wrapper.

The British Oil bottle was tall and slender and it rested on a square base. Godfrey's Cordial came in a conical vial with steep-pitched sides, the cone's point replaced by a narrow mouth.[55] Bateman's Pectoral Drops were packaged in a more common "phial"—a tall and slender cylindrical bottle.[56] Dalby's Carminative came in a bottle not unlike the Godfrey's Cordial bottle, except that Dalby's was impressed with the inscription dalby's carminativ.[57] Steer's Opodeldoc bottles were cylindrical in shape, with a wide mouth; some apparently were inscribed opodeldoc while others carried no such inscription. At least one brand of Daffy's Elixir was packaged in a globular bottle, according to a picture in a 1743 advertisement.[58] Speculation regarding the size and shape of the Stoughton bottle varies.[59] At least one Stoughton bottle was described as "Round amber. Tapered from domed shoulder to base. Long 5 in. bulged neck. Square flanged mouth. Flat base."[60]

Hooper's and Anderson's Scots Pills were, of course, not packaged in bottles (at least not the earliest), but were instead sold in the typical oval chip-wood pill boxes. On the lid of the box containing Hooper's Pills was stamped this inscription: dr. john hooper's female pills: by the king's patent 21 july 1743 no. 592. So far no example or illustration of Anderson's Scots Pills has been found. At least one producer, it will be remembered (page 157), sealed the box in black wax bearing a lion rampant, three mallets argent, and the bust of Dr. Anderson.

Source of Supply Severed

On September 29, 1774, John Boyd's "medicinal store" in Baltimore followed the time-honored custom of advertising in the *Maryland Gazette* a fresh supply of medicines newly at hand from England. To this intelligence was added a warning. Since nonimportation agreements by colonial merchants were imminent, which bade fair to make goods hard to get, customers would be wise to make their purchases before the supply became exhausted. Boyd's prediction was sound. The Boston Tea Party of the previous December had evoked from Parliament a handful of repressive measures, the

Intolerable Acts, and at the time of Boyd's advertisement, the first Continental Congress in session was soon to declare that all imports from Great Britain should be halted.

Figure 7.—Bottles of British Oil, 19th and early 20th century, from the Samuel Aker, David and George Kass collection, Albany, New York. (*Smithsonian photo 44201-B.*)

This Baltimore scare advertising may well have been heeded by Boyd's customers, for trade with the mother country had been interrupted before; in the wake of the Townshend Acts in 1767, when Parliament had placed import duties on various products, including tea, American merchants in various cities had entered into nonimportation agreements. Certainly, there was a decided decrease in the Boston advertising of patent medicines received from London. With respect to imports of any kind, it became necessary to explain, and one merchant noted that his goods were "the Remains of a Consignment receiv'd before the Non-Importation Agreement took place."[61] When Parliament yielded to the financial pressure and abolished all the taxes but the one on tea, nonimportation collapsed. This fact is reflected in an advertisement listing nearly a score of patent medicines, including the remedies of Turlington, Bateman, the Bettons, Anderson, Hooper, Godfrey, Daffy, and Stoughton, as "Just come to Hand and Warranted Genuine" on Captain Dane's ship, "directly from the Original Warehouse kept by dicey and okell in Bow Street, London."[62]

The days of such ample importations, however, were doomed, as commerce fell prey to the growing revolutionary agitation. The last medical advertisement in the *Massachusetts Gazette and Boston Weekly News-Letter*, before its demise the following February, appeared five months after the Battles of Lexington and Concord.[63] The apothecary at the Sign of the Unicorn was frank about the situation. He had imported fresh drugs and medicines every fall and spring up to the preceding June. He still had some on hand. Doctors and others should be advised.

Implicit in the advertisement is the suggestion that the securing of new supplies under the circumstances would be highly uncertain. That pre-war stocks did hold out, sometimes well into the war years may be deduced from a Williamsburg apothecary's advertisement.[64] W. Carter took the occasion of the ending of a partnership with his brother to publish a sort of inventory. Along with the "syrup and ointment pots, all neatly painted and lettered," the crabs eyes and claws, the Spanish flies, he listed a dozen patent medicines, including the remedies of Anderson, Bateman, and Daffy.

Even the British blockade failed to prevent patent medicines from being shipped from wholesaler to retailer. In the account book of a Salem, Massachusetts, apothecary,[65] the following entry appears:

4 cases Containing	
1 Dozn Bottles Godfreys Cordial	4/
5 Dozn Do Smaller Turling Bals	18/
8 Dozn Bettons British Oil	8/
6½ Dozn Hoopers Female Pills	10/
4 Dozn nd 8 Boxs And. Pills	10/

SALEM APRIL 8th 1777

The above 13 packages and 4 cases of medicines are ship'd on Board the Sloop Called the Two Brothers Saml West Master. On Account and [illegible word] of Mr. Oliver Smith of Boston Apothecary and to him consigned. The cases are unmarked being ship'd at Night. Error Excepted Jon. Waldo.

Figure 8.—Dalby's Carminative, two sides of a bottle from the McKearin collection, Hoosick Falls, New York. (*Smithsonian photo 44287-C.*)

The sloop was undoubtedly one of the small coastal type ships employed by the colonists, and the British blockade required such ominous precautions as "unmarked cases" and "ship'd by Night."

Such random assortments of prewar importations could hardly have met the American demand for the old English patent medicines created by a half century of use. Doubtless many embattled farmers had to confront their ailments without the accustomed English-made remedies. However, as early as the 1750's, at least two of the English patent medicines, Daffy's and Stoughton's Elixirs, were being compounded in the colonies and packaged in empty bottles shipped from England. Apothecary Carter of Williamsburg ordered sizable quantities of empty "Stoughton Vials" from 1752 through 1770, and occasionally ordered empty Daffy's bottles.[66] In 1774 apothecary Waldo of Salem noted the receipt from England of "1 Groce Stoughton Phials" and "1 Groce Daffy's Do."[67] Joseph Stansbury, who sold china and glass in Philadelphia, advertised "Daffy's Elixir Bottles" a week after the Declaration of Independence.[68] Stoughton's and Daffy's Elixirs, therefore, were being compounded by the American apothecaries during the Revolutionary War. Formulas for both preparations were official in the London and Edinburgh pharmacopoeias, as well as in unofficial formularies like Quincy's *Pharmacopoeias officinalis extemporanea* of 1765. All these publications were used widely by American physicians and apothecaries.

It is not known how extensively, during the struggle for independence, this custom was adopted for English patent medicines other than Daffy's and Stoughton's. However, imitation of English patent medicines in America was to increase, and it contributed to the chaos that beset the nostrum field when the war was over and the original articles from England were once more available. And they were bought. An advertisement at a time when the fighting was over and peace negotiations were still under way indicated that the Baltimore post office had half a dozen of the familiar English remedies for sale.[69] Two years later a New York store turned to tortured rhyme to convey the same message:[70]

> Medicines approv'd by royal charter,
>
> James, Godfry, Anderson, Court-plaster,

> With Keyser's, Hooper's Lockyer's Pills,
> And Honey Balsam Doctor Hill's;
> Bateman and Daffy, Jesuits drops,
> And all the Tinctures of the shops,
> As Stoughton, Turlington and Grenough,
> Pure British Oil and Haerlem Ditto....

Later in the decade, the Salem apothecary, Jonathon Waldo, made a list of "An assortment [of patent medicines] Usually Called For." The imported brand of Turlington's Balsam, Waldo stated, was "very dear" at 36 shillings a dozen, adding that his "own" was worth but 15 shillings for the same quantity. The English original of another nostrum, Essence of Peppermint, he listed at 18 shillings a dozen, his own at a mere 10/6.[71] Despite the price differential, importations continued. A Beverly, Massachusetts, druggist, Robert Rantoul, in 1799 ordered from London filled boxes and bottles of Anderson's Pills, Bateman's Drops, Steer's Opodeldoc, and Turlington's Balsam, along with the empty vials in which to put British Oil and Essence of Peppermint.[72] For decades thereafter the catalogs of wholesale drug firms continued to specify two grades of various patent medicines for sale, termed "English" and "American," "true" and "common," or "genuine" and "imitation."[73] This had not been the case in patent medicine listings of 18th-century catalogs.[74]

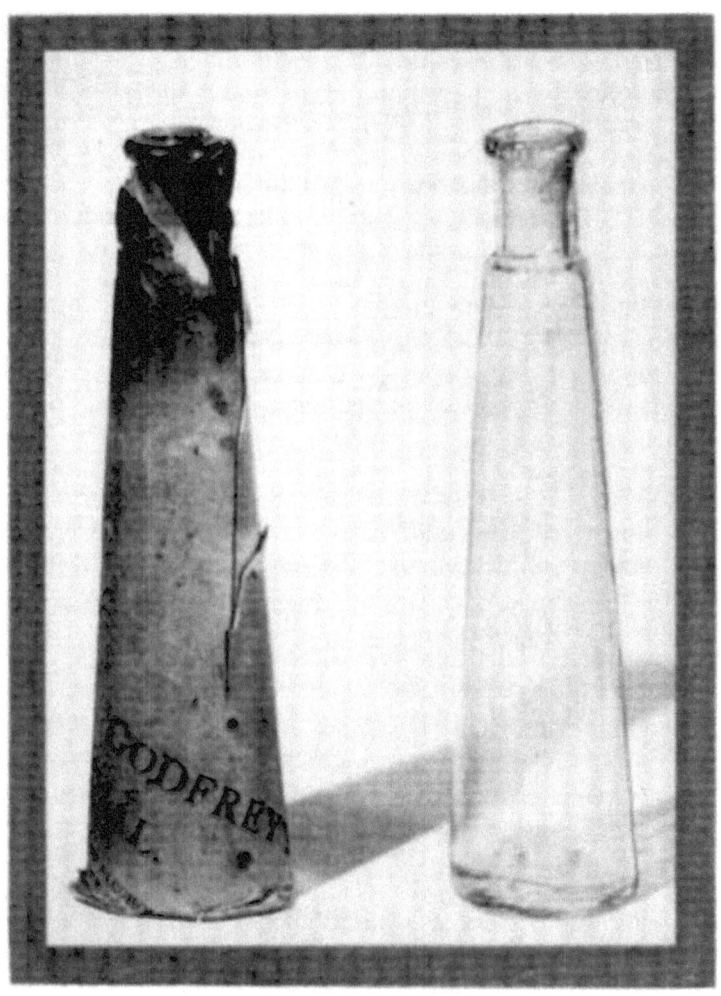

Figure 9.—Godfrey's Cordial, 19th-century bottles from the Samuel Aker, David and George Kass collection, Albany, New York. (*Smithsonian photo 44201-C.*)

In buying Anderson's and Bateman's remedies from London in 1799, Robert Rantoul of Massachusetts specified that they be secured from Dicey. It will be remembered that 60 years earlier William Dicey, John Cluer, and Robert Raikes were the group of entrepreneurs who had aided Benjamin Okell in patenting the pectoral drops bearing Bateman's name. Then and throughout the century, this concern continued to operate a warehouse in the Bow Churchyard, Cheapside, London. In 1721, it was known as the "Printinghouse and Picture Warehouse" of John Cluer, printer,[75] but by 1790, it was simply the "Medicinal Warehouse" of Bow Churchyard, Cheapside. This address lay in the center of the London area whence came nearly all of the British goods exported to America.[76] It had been the location of many merchants who had migrated to New England in the 17th century, and these newcomers had done business with their erstwhile associates who did not leave home. Thus were started trade channels which continued to run. The Bow Churchyard Warehouse may have been the major exporter of English patent medicines to colonial America, although others of importance were located in the same London region, in particular Robert Turlington of Lombard Street and Francis Newbery of St. Paul's Churchyard. The significance of the fact that there were key suppliers of patent medicines for the American market lies in the selection process which resulted. Out of the several hundred patent medicines which 18th-century Britain had available, Americans dosed themselves with that score or more which the major exporters shipped to colonial ports.

Not only did the Bow Churchyard Warehouse firm have Bateman's Drops. It will be remembered that in 1721 they advertised that they were preparing Daffy's Elixir. In 1743, they and Newbery were made exclusive vendors of Hooper's Pills.[77] By 1750, the firm was also marketing British Oil, Anderson's Pills, and Stoughton's Elixir.[78] Turlington in 1755 was selling not only his Balsam of Life, but was also vending Daffy's Elixir, Godfrey's Cordial, and Stoughton's Elixir.[79] After the tension of the Townshend Acts, it was the Bow Churchyard Warehouse which supplied a Boston apothecary with a large supply of nostrums, including all the eight patent medicines then in existence of the ten with which this discussion is primarily concerned.[80] On November 29, 1770, the *Virginia Gazette* (edited by Purdie and Dixon) reported a shipment, including Bateman's, Hooper's, Betton's, Anderson's, and Godfrey's remedies, just received "from Dr. Bateman's original wholesale warehouse in London" (the Bow Churchyard Warehouse). When Dalby's Carminative and Steer's Opodeldoc came on the market in the 1780's, it was Francis Newbery who had them for sale. Both the Newbery and Dicey (Bow Churchyard Warehouse) firms continued to operate in the post-Revolutionary years. Thus, it was no accident but rather vigorous commercial promotion over the decades, that resulted in the most popular items on the Dicey and Newbery lists appearing in the Philadelphia College of Pharmacy pamphlet published in 1824. And although the same old firms continued to export the same old medicines to the new United States, the back of the business was broken. The imitation spurred by wartime necessity became the post-war pattern.

The key recipes were to be found in formula books. Beginning in the 1790's, even American editions of John Wesley's *Primitive physic* included formulas for Daffy's, Turlington's, and Stoughton's remedies which the founder of Methodism had introduced into English editions of this guidebook to health shortly before his death.[81]

The homemade versions, as Jonathon Waldo had recorded (see p. 171), were about half as costly. The state of affairs at the turn of the new century is illustrated in the surviving business papers of the Beverly druggist, Robert Rantoul. In 1799 he had imported the British Oil and Essence of Peppermint bottles. In 1802 he reordered the latter, specifying that they should not have molded in the glass the words "by the Kings Patent." Rantoul wrote a formula for this nostrum in his formula book, and from it he filled 66 bottles in December 1801 and 202 bottles in June 1803. About the same time he began making and bottling Turlington's Balsam, ordering bottles of two sizes from London. His formula book contains these entries: "Jany 4th, 1804 filled 54 small turlingtons with 37 oz. Balsam," and "Jany 20th, 1804 filled 144 small turlingtons with 90¼ oz. Balsam and 9 Large Bottles with 8¼ oz."[82]

Two decades later the imitation of the English proprietaries was even bigger business. In 1821 William A. Brewer became apprenticed to a druggist in Boston. A number of the old English brands, he recalled, were still imported and sold at the time. But his apprenticeship years were heavily encumbered with duties involving the American versions. "Many, very many, days were spent," Brewer remembered, "in compounding these imitations, cleaning the vials, fitting, corking, labelling, stamping with fac-similes of the English Government stamp, and in wrapping them, with ... little regard to the originator's rights, or that of their heirs...." The British nostrums chiefly imitated in this Boston shop were Steer's, Bateman's, Godfrey's, Dalby's, Betton's, and Stoughton's. The last was a major seller. The store loft was mostly filled with orange peel and gentian, and the laboratory had "a heavy oaken press, fastened to the wall with iron clamps and bolts, which was used in pressing out 'Stoughton's Bitters,' of which we usually prepared a hogshead full at one time." A large quantity was needed. In those days, Brewer asserted, "almost everybody indulged in Stoughton's elixir as morning bitters." [83]

Figure 10.—Godfrey's Cordial, early 20th century bottles manufactured in the U.S.A. (*U.S. National Museum cat. Nos. M-6989, and M-6990; Smithsonian photo 44287 B.*)

Other drugstores certainly followed the practice of Brewer's employer, in cleaning up and refilling bottles that had previously been drained of their old English medicines. The chief source of bottles to hold the American imitations, however, was the same as that to which Waldo and Rantoul had turned, English glass factories. It was not so easy for Americans to fabricate the vials as it was for them to compound the mixtures to fill them. In the years before the War of 1812, the British glass industry maintained a virtual monopoly of the specially-shaped bottles for Bateman's, Turlington's, and the other British remedies. When in the 1820's the first titan of made-in-America nostrums, Thomas W. Dyott of Philadelphia, appeared upon the scene, this venturesome entrepreneur decided to make bottles not only for his own assorted remedies but also for the popular English brands. In time he succeeded in improving the quality of American bottle glass and in drastically reducing prices. The standard cost for most of the old English vials under the British monopoly had been $5.50 a gross. By the early 1830's Dyott had cut the price to under two dollars.[84]

Figure 11.—An Original Package of Hooper's Pills, from the Samuel Aker, David and George Kass collection, Albany, New York. (*Smithsonian photo* 44201.)

Other American glass manufactories followed suit. For example, in 1835 the Free Will Glass Manufactory was making "Godfrey's Cordial," "Turlington's Balsam," and "Opodeldoc Bitters bottles."[85] An 1848 broadside entitled "The Glassblowers' List of Prices of Druggist's Ware," a broadside preserved at the Smithsonian Institution, includes listings for Turlington's Balsam, Godfrey's Cordial,

Dalby's and Small and Large Opodeldoc bottles, among many other American patent medicine bottles.

In the daybook of the Beverly, Massachusetts, apothecary,[86] were inscribed for Turlington's Balsam, three separate formulas, each markedly different from the others. A Philadelphia medical journal in 1811 contained a complaint that Americans were using calomel in the preparation of Anderson's Scots Pills, and that this practice was a deviation both from the original formula and from the different but still all-vegetable formula by which the pills were being made in England.[87] Various books were published revealing the "true" formulas, in conflicting versions.[88]

Philadelphia College of Pharmacy Formulary

As the years went by and therapeutic laissez-faire continued to operate, conditions worsened. By the early 1820's, the old English patent medicines, whether of dwindling British vintage or of burgeoning American manufacture, were as familiar as laudanum or castor oil.

With the demand so extensive and the state of production so chaotic, the officials of the new Philadelphia College of Pharmacy were persuaded that remedial action was mandatory. In May 1822, the Board of Trustees resolved to appoint a 5-man committee "to select from such prescriptions for the preparation of Patent Medicines ..., as may be submitted to them by the members of the College, those which in their opinion, may be deemed most appropriate for the different compositions."

The committee chose for study "eight of the Patent Medicines most in use," and sought to ascertain what ingredients these ancient remedies ought by right to contain. Turning to the original formulas, where these were given in English patent specifications, the pharmacists soon became convinced that the information provided by the original proprietors served "only to mislead."

If the patent specifications were perhaps intentionally confusing, the committee inquired, how could the original formulas really be known? This quest seemed so fruitless that it was not pursued. Instead the pharmacists turned to American experience in making the

English medicines. From many members of the College, and from other pharmacists as well, recipes were secured. The result was shocking. Although almost every one came bolstered with the assertion that it was true and genuine, the formulas differed so markedly one from the other, the committee reported, as to make "the task of reformation a very difficult one." Indeed, in some cases, when two recipes bearing the same old English name were compared, they were found to contain not one ingredient in common. In other cases, the proportions of some basic ingredient would vary widely. All the formulas collected for Bateman's Pectoral Drops, for instance, contained opium, but the amount of opium to liquid ingredients in one formula submitted was 1 to 14, while in another it was 1 to 1,000.

Setting forth boldly to strip these English nostrums of "their extravagant pretensions," the committee sought to devise formulas for their composition as simple and inexpensive as possible while yet retaining the "chief compatible virtues" ascribed to them on the traditional wrappers.

Hooper's Female Pills had been from the beginning a cathartic and emmenagogue. However, only aloes was common to all the recipes submitted to the committee. This botanical, which still finds a place in laxative products today, was retained by the committee as the cathartic base, and to it were added "the Extract of Hellebore, the Sulphate of Iron and the Myrrh as the best emmenagogues."

Anderson's Scots Pills had been a "mild" purgative throughout its long career, varying in composition "according to the judgement or fancy of the preparer." Paris, an English physician, had earlier reported that these pills consisted of aloes and jalap; the committee decided on aloes, with small amounts of colocynth and gamboge, as the purgatives of choice.

Of Bateman's Pectoral Drops more divergent versions existed than of any of the others. The committee settled on a formula of opium and camphor, not unlike paragoric in composition, with catachu, anise flavoring, and coloring added. Godfrey's Cordial also featured opium in widely varying amounts. The committee chose a formula which would provide a grain of opium per ounce, to which was added sassafras "as the carminative which has become one of the chief features of the medicine."

English apothecary Dalby had introduced his "Carminative" for "all those fatal Disorders in the Bowels of Infants." The committee decided that a grain of opium to the ounce, together with magnesia and three volatile oils, were essential "for this mild carminative and laxative ... for children."

Instead of the complex formula described by Robert Turlington for his Balsam of Life, the committee settled on the official formula of Compound Tincture of Benzoin, with balsam of peru, myrrh, and angelica root added, to produce "an elegant and rich balsamic tincture." On the other hand, the committee adopted "with slight variations, the Linimentum Saponis of the old London Dispensatory" to which they, like Steers, added only ammonia.

The committee found two distinct types of British Oil on the market. One employed oil of turpentine as its basic ingredient, while the other utilized flaxseed oil. The committee decided that both oils, along with several others in lesser quantities, were necessary to produce a medicine "as exhibited in the directions" sold with British Oil. "Oil of Bricks" which apparently was the essential ingredient of the Betton British Oil, was described by the committee as "a nauseous and unskilful preparation, which has long since been banished from the Pharmacopoeias."

Thus the Philadelphia pharmacists devised eight new standardized formulas, aimed at retaining the therapeutic goals of the original patent medicines, while brought abreast of current pharmaceutical knowledge. Recognizing that the labeling had long contained "extravagant pretensions and false assertions," the committee recommended that the wrappers be modified to present only truthful claims. If the College trustees should adopt the changes suggested, the committee concluded optimistically, then "the reputation of the College preparations would soon become widely spread, and we ... should reap the benefit of the examination which has now been made, in an increased public confidence in the Institution and its members; the influence of which would be felt in extending the drug business of our city."[89]

The trustees felt this counsel to be wise, and ordered 250 copies of the 12-page pamphlet to be printed. So popular did this first major undertaking of the Philadelphia College prove that in 1833 the formulas were reprinted in the pages of the journal published by the College.[90] Again the demand was high, few numbers of the publication were "more sought after," and in 1839 the formulas were printed once again, this time with slight revisions.[91]

Thus had the old English patent medicines reached a new point in their American odyssey. They had first crossed the Atlantic to serve the financial interests of the men who promoted them. During the Revolution they had lost their British identity while retaining their British names. The Philadelphia pharmacists, while adopting them and reforming their character, did not seek to monopolize them, as had the original proprietors. They now could work for every man.

English Patent Medicines Go West

The double reprinting of the formulas was one token of the continuing role in American therapy of the old English patent medicines. There were others. In 1829 with the establishment of a school of pharmacy in New York City, the Philadelphia formulas were accepted as standard. The new labels devised by the Philadelphians with their more modest claims of efficacy had a good sale.[92] It was doubtless the Philadelphia recipes which went into the Bateman and Turlington and Godfrey vials with which a new druggist should be equipped "at the outset of business," according to a book of practical counsel.[93] To local merchants who lacked the knowledge or time to do it themselves, drummers and peddlers vended the medicines already bottled. "Doctor" William Euen of Philadelphia issued a pamphlet in 1840 to introduce his son to "Physicians and Country Merchants." His primary concern was dispensing nostrums bearing his own label, but his son was also prepared to take orders for the old English patent medicines.[94] Manufacturers and wholesalers of much better repute were prepared to sell bottles for the same brands, empty or filled.

Figure 12.—English and American Brands of Hooper's Female Pills, an assortment of packages of from the Samuel Aker, David and George Kass collection, Albany, New York. (*Smithsonian photo 44201-D.*)

In the early 1850's a young pharmacist in upstate New York,[95] using "old alcohol barrels for tanks," worked hard at concocting Bateman's and Godfrey's and Steer's remedies. John Uri Lloyd of Cincinnati recalled having compounded Godfrey's Cordial and Bateman's Drops, usually making ten gallons in a single batch.[96] Out in Wisconsin, another druggist was buying Godfrey's Cordial bottles at a dollar for half a gross, sticking printed directions on them that cost twelve cents for the same quantity, and selling the medicine at four ounces for a quarter.[97] He also sold British Oil and Opodeldoc, the same old English names dispensed by a druggist in another Wisconsin town, who in addition kept Bateman's Oil in stock at thirteen cents the bottle.[98] Godfrey's was listed in the 1860 inventory of an Illinois general store at six cents a bottle.[99]

Farther west the same familiar names appeared. Indeed, the old English patent medicines had long since moved westward with fur trader and settler. As early as 1783, a trader in western Canada, shot by a rival, called for Turlington's Balsam to stop the bleeding. Alas, in this case, the remedy failed to work.[100] In 1800 that inveterate Methodist traveler, Bishop Francis Asbury, resorted to Stoughton's Elixir when afflicted with an intestinal complaint.[101] In 1808, some two months after the first newspaper began publishing west of the Mississippi River, a local store advised readers in the vicinity of St. Louis that "a large supply of patent medicines" had just been received, among them Godfrey's Cordial, British Oil, Turlington's Balsam, and Steer's "Ofodeldo [sic]."[102]

Turlington's product played a particular role in the Indian trade, thus demonstrating that the red man has not been limited in nostrum history to providing medical secrets for the white man to exploit. Proof of this has been demonstrated by archaeologists working under the auspices of the Smithsonian Institution in both North and South Dakota. Two pear-shaped bottles with Turlington's name and patent claims embossed in the glass were excavated by a Smithsonian Institution River Basin Surveys expedition in 1952, on the site of an old trading post known as Fort Atkinson or Fort Bethold II, situated some 16 miles southeast of the present Elbowoods, North Dakota. In 1954 the North Dakota Historical Society found a third bottle nearby. These posts, operated from the mid-1850's to the mid-1880's, served the Hidatsa and Mandan Indians who dwelt in a town named Like-a-Fishhook Village. The medicine bottles were made of cast glass, light green in color, probably of American manufacture. More interesting is the bottle from South Dakota. It was excavated in 1923 near Mobridge at a site which was the principal village of the Arikara Indians from about 1800 to 1833, a town visited by Lewis and Clark as they ascended the Missouri River in 1804. This bottle, made of English lead glass and therefore an imported article, was unearthed from a grave in the Indian burying ground. Throughout history the claims made in behalf of patent medicines have been extreme. This Turlington bottle, however, affords one of the few cases on record wherein such a medicine has been felt to possess a postmortem utility.[103]

Fur traders were still using old English patent medicines at mid-century. Four dozen bottles of Turlington's Balsam were included in an "Inventory of Stock the property of Pierre Chouteau, Jr. and Co. U[pper], M[issouri]. On hand at Fort Benton 4th May 1851...."[104] In the very same year, out in the new State of California, one of the early San Francisco papers listed Stoughton's Bitters as among the merchandise for sale at a general store.[105]

Newspaper advertising of the English proprietaries—even the mere listing so common during the late colonial years—became very rare after the Philadelphia College of Pharmacy pamphlet was issued. Apothecary George J. Fischer of Frederick, Maryland, might mention seven of the old familiar names in 1837,[106] and another druggist in the same city might present a shorter list in 1844,[107] but such advertising was largely gratuitous. Since the English patent medicines had become every druggist's property, people who felt the need of such dosage would expect every druggist to have them in stock. There was no more need to advertise them than there was to advertise laudanum or leeches or castor oil. Even the Supreme Court of Massachusetts in 1837 took judicial cognizance of the fact that the old English patent medicine names had acquired a generic meaning descriptive of a general class of medicines, names which everyone was free to use and no one could monopolize.[108]

As the years went by, and as advertising did not keep the names of the old English medicines before the eyes of customers, it is a safe assumption that their use declined. Losing their original proprietary status, they were playing a different role. New American proprietaries had stolen the appeal and usurped the function which Bateman's Drops and Turlington's Balsam had possessed in 18th-century London and Boston and Williamsburg. As part of the cultural nationalism that had accompanied the Revolution, American brands of nostrums had come upon the scene, promoted with all the vigor and cleverness once bestowed in English but not in colonial American advertising upon Dalby's Carminative and others of its kind. While these English names retreated from American advertising during the 19th century, vast blocks of space in the ever-larger newspapers were devoted to extolling the merits of Dyott's Patent Itch Ointment, Swaim's Panacea, and Brandreth's Pills. More and more Americans were learning how to read, as free public education spread. Persuaded by the frightening symptoms and the glorious promises, citizens with a bent toward self-dosage flocked to buy the American brands. Druggists and general stores stocked them and made fine profits.[109] While bottles of British Oil sold two for a quarter in 1885 Wisconsin, one bottle of Jayne's Expectorant retailed for a dollar.[110] It is no wonder that, although the old English names continue to appear in the mid-19th-century and later druggists' catalogs and price currents,[111] they are muscled aside by the multitude of brash American nostrums. Many of the late 19th century listings continued to follow the procedure set early in the century of specifying two grades of the various patent medicines, *i.e.*, "English" and "American," "genuine" and "imitation," "U.S." and "stamped." American manufactories specializing in pharmaceutical glassware continued to offer the various English patent medicine bottles until the close of the century.[112]

Figure 13.—Opodeldoc Bottle from the collection of Mrs. Leo F. Redden, Kenmore, New York. (*Smithsonian photo 44201-E.*)

In a thesaurus published in 1899, Godfrey's, Bateman's, Turlington's, and other of the old English patent remedies were termed "extinct patents."[113] The adjective referred to the status of the patent, not the condition of the medicines. If less prominent than in the olden days, the medicines were still alive. The first edition of the *National Formulary*, published in 1888, had cited the old English names as synonyms for official preparations in four cases, Dalby's, Bateman's, Godfrey's and Turlington's.

Figure 14. — Opodeldoc Bottle as illustrated in the 1879 Catalog of Hagerty Bros., New York City, New York.

Thus as the present century opened, the old English patent medicines were still being sold. City druggists were dispensing them over their counters, and the peddler's wagon carried them to remote rural regions.[114] But the medical scene was changing rapidly. Improvements in medical science, stemming in part from the establishment of the germ theory of disease, were providing a better yardstick against which to measure the therapeutic efficiency of proprietary remedies. Medical ethics were likewise advancing, and the occasional critic among the ranks of physicians was being joined by scores of his fellow practitioners in lambasting the brazen effrontery of the hundreds of American cure-alls which advertised from newspaper and roadside sign. Journalists joined doctors in condemning nostrums. Samuel Hopkins Adams in particular, writing "The Great American Fraud" series for *Collier's Weekly*, frightened and aroused the American public with his exposure of cheap whiskey posing as consumption cures and soothing syrups filled with opium. Then came a revolution in public policy. After a long and frustrating legislative prelude, Congress in June of 1906 passed, and President Theodore Roosevelt signed, the first Pure Food and Drugs Act. The law contained clauses aimed at curtailing the worst features of the patent medicine evil.

The Patent Medicines In The 20th Century

Although the old English patent medicines had not been the target at which disturbed physicians and "muck-raking" journalists had taken aim, these ancient remedies were governed by provisions of the new law. In November 1906 the Bureau of Chemistry of the Department of Agriculture, in charge of administering the new federal statute, received a letter from a wholesale druggist in Evansville, Indiana. One of his stocks in trade, the druggist wrote, was a remedy called Godfrey's Cordial. He realized that the Pure Food and Drugs Act had something to do with the labeling of medicines containing opium, as Godfrey's did, and he wanted to know from the Bureau just what was required of him.[115] Many manufacturing druggists and producers of medicine were equally anxious to learn how the law would affect them. The editors of a trade paper, the *American Druggist and Pharmaceutical Record*, issued warnings and gave advice. It was still the custom, they noted, to wrap bottles of ancient patent medicines, like Godfrey's Cordial and Turlington's Balsam, in facsimiles of the original circulars, on which were printed extravagant claims and fabulous certificates of cures that dated back some two hundred years. The new law was not going to permit the continuation of such 18th-century practices. Statements on the label "false or misleading in any particular" were banned.[116]

A few manufacturers, as the years went by, fell afoul of this and other provisions of the law. In 1918 a Reading, Pennsylvania, firm entered a plea of guilty and received a fifty dollar fine for putting on the market an adulterated and misbranded version of Dr. Bateman's Pectoral Drops.[117] The law required that all medicines sold under a name recognized in the *United States pharmacopoeia* or the *National formulary*, and Bateman's was included in the latter, must not differ from the standard of strength, quality, or purity as established by these volumes. Yet the Bateman Drops produced in Reading, the government charged, fell short. They contained only 27.8 percent of the alcohol and less than a tenth of the morphine that they should have had. While short on active ingredients, the Drops were long on claims. The wrapper boasted that the medicine was "effective as a remedy for all fluxes, spitting of blood, agues, measles, colds, coughs, and to put off the most violent fever; as a treatment, remedy, and cure for stone and gravel in the kidneys, bladder, and urethra, shortness of breath, straightness of the breast; and to rekindle the most natural heat in the bodies by which they restore the languishing to perfect health." Okell and Dicey had scarcely promised more. By 20th-century standards, the government asserted, these claims were false and fraudulent.

Other manufacturers sold Bateman's Drops without running afoul of the law. In 1925, ninety-nine years after the Philadelphia College of Pharmacy pamphlet was printed, one North Carolina firm was persuaded that it still was relevant to tell potential customers, in a handbill, that its Drops were being made in strict conformity with the College formula.[118] For Compound Tincture of Opium and Gambir Compound, however, most manufacturers chose to follow the *National formulary* specifications, which remained official until 1936.

Another old English patent medicine against which the Department of Agriculture was forced to take action was Hooper's Female Pills. Between 1919 and 1923, government agents seized a great many shipments of this ancient remedy in versions put out by three Philadelphia concerns.[119] Some of the packages bore red seals, others green seals, and still others black, but the labeling of all claimed them to be "a safe and sovereign remedy in female complaints." This theme was expanded in considerable detail and there was an 18th-century ring to the promise that the pills would work a sure cure "in all hypochondriac, hysterick or vapourish disorders." No pill made essentially of aloes and ferrous sulphate, said the government experts, could do these things. Nor did the manufacturers, in court, seek to say otherwise. Whether the seals were green or red, whether the packages were seized in Washington or Worcester, the result was the same. No party appeared in court to claim the pills, and they were condemned and destroyed.

In one of the last actions under the 1906 law, a case concluded in 1940, after the first federal statute had been superseded by a more rigorous one enacted in 1938, two of the old English patent medicines encountered trouble.[120] They were British Oil and Dalby's Carminative, as prepared by the South Carolina branch of a large pharmaceutical manufacturing concern.

According to the label, the British Oil was made in conformity with the Philadelphia College of Pharmacy formula given in an outdated edition of the *United States dispensatory*. But instead of containing a proper amount of linseed oil, if indeed it contained any, the medicine was made with cottonseed oil, an ingredient not mentioned in the Dispensatory. Therefore, the government charged, the Oil was adulterated, under that provision of the law requiring a medicine to maintain the strength and purity of any standard it professed to follow. More than that, the labeling contravened the law since it represented the remedy as an effective treatment for various swellings, inflammations, fresh wounds, earaches, shortnesses of breath, and ulcers.

Dalby's Carminative was merely misbranded, but that was bad enough. Its label suggested that it be used especially "For Infants Afflicted With Wind, Watery Gripes, Fluxes and Other Disorders of the Stomach and Bowels," although it would aid adults as well. The impression that this remedy was capable of curing such afflictions, the government charged, was false and fraudulent. Moreover, since the Carminative contained opium, it was not a safe medicine when given according to the dosage directions in a circular accompanying the bottle. For these and several other violations of the law, the defending company, which did not contest the case, was fined a hundred dollars.

Throughout the 19th century, occasional criticism of the old English patent medicines had been made in the lay press. One novel[121] describes a physician who comments on the use of Dalby's Carminative for babies: "Don't, for pity's sake, vitiate and torment your poor little angel's stomach, so new to the atrocities of this world, with drugs. These mixers of baby medicines ought to be fed nothing but their own nostrums. That would put a stop to their inventions of the adversary."

Opium had been lauded in the 17th and 18th centuries, when the old English proprietaries began, as a superior cordial which could moderate most illnesses and even cure some. "Medicine would be a one-armed man if it did not possess this remedy." So had stated the noted English physician, Thomas Sydenham.[122] But the 20th century had grown to fear this powerful narcotic, especially in remedies for children. This point of view, illustrated in the governmental action concerning Dalby's Carminative, was also reflected in medical comment about Godfrey's Cordial. During 1912, a Missouri physician described the death of a baby who had been given this medicine for a week.[123] The symptoms were those of opium poisoning. Deploring the naming of this "dangerous mixture" a "cordial," since the average person thought of a cordial as beneficial, the doctor hoped that the formula might be omitted from the next edition of the *National formulary*. This did not happen, for the recipe hung on until 1926. The Harrison Narcotic Act, enacted in 1914 as a Federal measure to restrict the distribution of narcotics,[124] failed to restrict the sale of many opium-bearing compounds like Godfrey's Cordial. In 1931, a Tennessee resident complained to the medical journal *Hygeia* that this medication was "sold in general stores and drug stores here without prescription and is given to babies." To this, the journal replied that the situation was "little short of criminal."[125] The charge leveled against his competitors by one of the first producers of Godfrey's Cordial two centuries earlier (see page 158) may well have proved a prophecy broad enough to cover the whole history of this potent nostrum. "... Many Men, Women, and especially Infants," he said, "may fall as Victims, whose Slain may exceed Herod's Cruelty...."

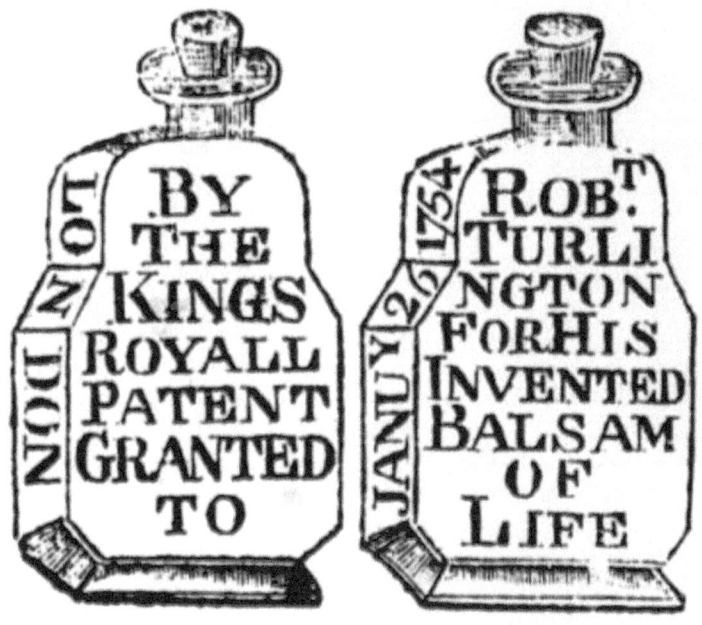

Figure 15.—Turlington's Balsam of Life bottles as pictured in a brochure dated 1755-1757, preserved in the Pennsylvania Historical Society, Philadelphia, Pa. According to Turlington, the bottle was adopted in 1754 "to prevent the villainy of some persons who, buying up my empty bottles, have basely and wickedly put therein a vile spurious counterfeit sort."

For those who persist in using the formulas of the early English patent medicines, recipes are still available. Turlington's Balsam remains as an unofficial synonym of U.S.P. Compound Tincture of Benzoin. Concerning its efficacy, the *United States dispensatory*[126] states: "The tincture is occasionally employed internally as a stimulating expectorant in chronic bronchitis. More frequently it is used as an inhalent ... It has also been recommended in chronic dysentery ... but is of doubtful utility."

A formula for Godfrey's Cordial, under the title of Mixture of Opium and Sassafras, is still carried in the *Pharmaceutical recipe book*.[127] *Remington's practice of pharmacy*[128] retains a formula for Dalby's Carminative under the former *National formulary* title of Carminative Mixture.

In the nation of their origin, the continuing interest in the ancient proprietaries seems somewhat more lively than in America. The 1953 edition of *Pharmaceutical formulas,* published by the London journal *The Chemist and Druggist,* includes formulas for eight of the ten old patent medicines described in this study. This compendium, indeed, lists not one, but three different recipes for British Oil, and the formulas by which Dalby's Carminative may be compounded run on to a total of eight. Two lineal descendents of 18th-century firms which took the lead in exporting to America still manufacture remedies made so long ago by their predecessors. May, Roberts & Co., Ltd., of London, successors to the Newbery interests, continues to market Hooper's Female Pills, whereas W. Sutton & Co. (Druggists' Sundries), London, Ltd., of Enfield, in Middlesex, successors to Dicey & Co. at Bow Churchyard, currently sells Bateman's Pectoral Drops.[129]

In America, however, the impact of the old English patent medicines has been largely absorbed and forgotten. During the past twenty years a revolution in medical therapy has taken place. Most of the drugs in use today were unknown a quarter of a century ago. Some of the newer drugs can really perform certain of the healing miracles claimed by their pretentious proprietors for the old English patent medicines.

A more recent import from Britain, penicillin, may prove to have an even longer life on these shores than did Turlington's Balsam or Bateman's Drops. Still, two hundred years is a long time. Despite the fact that these early English patent medicines are nearly forgotten by the public today, their American career is none the less worth tracing. It reflects aspects not only of medical and pharmaceutical history, but of colonial dependence, cultural nationalism, industrial development, and popular psychology. It reveals how desperate man has been when faced with the terrors of disease, how he has purchased the packaged promises offered by the sincere but deluded as well as by the charlatan. It shows how science and law have combined to offer man some safeguards against deception in his pursuit of health.

The time seems ripe to write the epitaph of the old English patent medicines in America. That they are now a chapter of history is a token of medical progress for mankind.

Figure 16.—Turlington's Balsam of Life Bottle (all four sides) found in an Indian grave at Mobridge, South Dakota; now preserved in the U.S. National Museum. (*Cat. No. 32462, Archeol.; Smithsonian photo 42936-A.*)

Footnotes

¹ Unless otherwise indicated, the early English history of these patent medicines has been obtained from the following sources: "Proprietaries of other days," *Chemist and Druggist*, June 25, 1927, vol. 106, pp. 831-840; C. J. S. Thompson, *The mystery and art of the apothecary*, London, 1929; C. J. S. Thompson, *Quacks of old London*, London, 1928; and A. C. Wootton, *Chronicles of pharmacy*, London, 1910, 2 vols.

² "How the patent medicine industry came into its own," *American Druggist*, October 1933, vol. 88, pp. 84-87, 232, 234, 236, 238.

³ Benjamin Okell, "Pectoral drops for rheumatism, gravel, etc.," British patent 483, March 31, 1726.

⁴ British Patent Office, *Patents for inventions: abridgements of specifications relating to medicine, surgery, and dentistry, 1620-1866*, London, 1872.

⁵ *London Mercury*, London, August 19-26, 1721.

⁶ *A short treatise of the virtues of Dr. Bateman's Pectoral Drops*, New York, 1731. A 36-page pamphlet preserved in the Library of the New York Academy of Medicine. This is an American reprint of an English original, date unknown.

⁷ A broadside, issued in London, *ca.* 1750, advertising "Dr. Bateman's Drops," is preserved in the Warshaw Collection of Business Americana, New York. Later reprints of this same broadside are preserved in the private collection of Samuel Aker, Albany, New York, and in the Smithsonian Institution.

[8] Michael and Thomas Betton, "Oil for the cure of rheumatic and scorbutic affections," British patent 587, August 14, 1742.

[9] Edmund Darby & Co., *Directions for taking inwardly and using outwardly the company's true genuine and original British Oil; prepared by Edmund Darby & Co. at Coalbrook-Dale, Shropshire*, ca. 1745. An 8-page pamphlet preserved in the Library of the College of Physicians, Philadelphia, Pennsylvania.

[10] *London Gazette*, London, March 1, 1745.

[11] John Hooper, "Pills," British patent 592, July 21, 1743.

[12] E. Burke Inlow, *The patent grant*, Baltimore, 1950, p. 33.

[13] *Daily Advertiser*, London, September 23, 1743.

[14] Robert Turlington, "A Specifick balsam, called the balsam of life," British patent 596, January 18, 1744.

[15] Robert Turlington, *Turlington's Balsam of Life*, ca. 1747. A 46-page pamphlet preserved in the Folger Shakespeare Library, Washington, D. C.

[16] *Daily Advertiser*, London, February 18, 1780.

[17] Broadsides, *ca.* 1810-1822, advertising "Steer's Chemical Opodeldoc, for bruises, sprains, rheumatism, etc., etc.," are preserved in the American Antiquarian Society, Worcester, Massachu-

setts; the Library of the New York Academy of Medicine; and the Warshaw Collection of Business Americana, New York.

[18] *Daily Advertiser*, London, January 4, 1781.

[19] *Ibid.*, September 7, 1743.

[20] *London Mercury*, London, August 19-26, 1721.

[21] Richard Stoughton, "Restorative cordial and medicine," British patent 390, 1712.

[22] From a broadside, *ca.* 1750, advertising "Dr. Stoughton's Elixir Magnum Stomachum," preserved in the American Antiquarian Society, Worcester, Mass.

[23] British Patent Office, *op. cit.* (see footnote 4).

[24] Poplicola, "Pharmacopoeia empirica or the list of nostrums and empirics," *The Gentleman's Magazine*, 1748, vol. 18, pp. 346-350.

[25] George L. Kittredge, "Letters to Samuel Lee and Samuel Sewall relating to New England and the Indians," *Colonial Society of Massachusetts, Transactions*, 1913, vol. 14, pp. 142-186.

[26] Bartholomew Brown, Apothecary day book, Salem [1698]; manuscript original preserved in the Library of the Essex Institute, Salem, Massachusetts.

[27] Frank L. Mott, *American journalism*, New York, 1941, pp. 9-10.

[28] *Boston News-Letter*, Boston, February 9, 1708.

[29] *Ibid.*, March 12, 1711.

[30] *Ibid.*, March 24, 1712.

[31] *Ibid.*, November 14, 1720.

[32] *Ibid.*, March 12, 1730.

[33] *Ibid.*, January 4, 1739.

[34] *Ibid.*, November 14, 1748.

[35] *Ibid.*, June 7, 1750.

[36] *Ibid.*, May 24, 1750.

[37] *Ibid.*, December 31, 1761.

[38] Lester J. Cappon and Stella F. Duff, *Virginia Gazette index, 1736-1780*, Williamsburg, 1950, 2 vols.

[39] *Virginia Gazette*, Williamsburg, May 27, 1737.

[40] *Pennsylvania Gazette*, Philadelphia, December 1, 1768.

[41] *Boston News-Letter*, Boston, November 24, 1763.

[42] James J. Walsh, *History of the Medical Society of the State of New York*, New York, 1907.

[43] Robert Turlington, "Turlington's Balsam of Life," 1755-1757. A later reprint of this same circular is preserved in the Warshaw Collection of Business Americana.

[44] *Turlington's Balsam of Life* (see footnote 15).

[45] "Dr. Bateman's Drops" (see footnote 7).

[46] James Carter, Apothecary account book, Williamsburg [1752-1773]. Manuscript original preserved at Colonial Williamsburg, Virginia.

[47] *A short treatise of the virtues of Dr. Bateman's Pectoral Drops* (see footnote 6).

[48] Gertrude L. Annan, "Printing and medicine," *Bulletin of the Medical Library Association*, March 1940, vol. 28, p. 155.

[49] Wyndham B. Blanton, *Medicine in Virginia in the eighteenth century*, Richmond, Virginia, 1931, pp. 33-34.

[50] Maurice Bear Gordon, *Aesculapius comes to the colonies*, Ventnor, New Jersey, 1949, p. 39.

[51] Fielding H. Garrison, *An introduction to the history of medicine*, Philadelphia, 1924, pp. 405-408; and Richard H. Shryock. *The development of modern medicine*, New York, 1947, pp. 51-54.

[52] Kittredge, *op. cit.* (footnote 25).

[53] "From past times an original bottle of Turlington's Balsam," *Chemist and Druggist*, September 23, 1905, vol. 67, p. 525; Stewart Schackne, "Bottles," *American Druggist*, October 1933, vol. 88, pp. 78-81, 186-188, 190, 194; Frederick Fairchild Sherman, "Some early bottles," *Antiques*, vol. 3, pp. 122-123; and Stephen Van Rensselaer, *Early American bottles and flasks*, Peterborough, New Hampshire, 1926.

[54] Waldo R. Wedel and George B. Griffenhagen, "An English balsam among the Dakota aborigines," *American Journal of Pharmacy*, December 1954, vol. 126, pp. 409-415.

[55] Sherman, *op. cit.* (footnote 53).

[56] Schackne, *op. cit.* (footnote 53).

[57] George S. and Helen McKearin, *American glass*, New York, 1941.

[58] *Daily Advertiser*, London, October 29, 1743.

[59] George Griffenhagen, "Stodgy as a Stoughton bottle," *Journal of the American Pharmaceutical Association, Practical Pharmacy Edition*, January 1956, vol. 17, p. 20; Mitford B. Mathews, ed., *A dictionary of Americanisms on historical principles*, Chicago, 1951, 2 vols.; Bertha Kitchell Whyte, *Wisconsin heritage*, Boston, 1954; Charles Earle Funk, *Heavens to Betsy! and other curious sayings*, New York, 1955.

[60] James H. Thompson, *Bitters bottles*, Watkins Glen, New York, 1947, p. 60.

[61] *Massachusetts Gazette*, Boston, December 21, 1769.

[62] *Ibid.*, April 25, 1771.

[63] *Ibid.*, September 7, 1775.

[64] *Virginia Gazette* (edited by Dixon and Nicholson), Williamsburg, June 12, 1779.

[65] Jonathon Waldo, Apothecary account book, Salem, Massachusetts [1770-1790]. Manuscript original preserved in the Library of the Essex Institute, Salem, Mass.

[66] Carter, *op. cit.* (footnote 46).

[67] Waldo, *op. cit.* (footnote 65).

[68] *Pennsylvania Gazette*, Philadelphia, July 11, 1776.

[69] *Maryland Journal and Baltimore Gazette*, Baltimore, October 29, 1782.

[70] *New York Packet and the American Advertiser*, New York, October 11, 1784.

⁷¹ Waldo, *op. cit.* (footnote 65).

⁷² Robert Rantoul, Apothecary daybooks, 3 vols., Beverly, Massachusetts [1796-1812]. Manuscript originals preserved in the Beverly Historical Society. Also see Robert W. Lovett, "Squire Rantoul and his drug store," *Bulletin of the Business Historical Society*, June 1951, vol. 25, pp. 99-114.

⁷³ Joel and Jotham Post, *A catalogue of drugs, medicines & chemicals, sold wholesale & retail, by Joel and Jotham Post, druggists, corner of Wall and William-Streets*, New York, 1804; Massachusetts College of Pharmacy, *Catalogue of the materia medica and of the pharmaceutical preparations, with the uniform prices of the Massachusetts College of Pharmacy*, Boston, 1828; George W. Carpenter, *Essays on some of the most important articles of the materia medica ... to which is added a catalogue of medicines, surgical instruments, etc.*, Philadelphia, 1834.

⁷⁴ John Dunlap, *Catalogus medicinarum et pharmacorum*, Philadelphia, 1771; John Day, *Catalogue of drugs, chymical and galenical preparations, shop furniture, patent medicines, and surgical instruments sold by John Day and Company, druggists and chymists in second-street*, Philadelphia, 1771; George Griffenhagen, "The Day-Dunlap 1771 pharmaceutical catalog," *American Journal of Pharmacy*, September 1955, vol. 127, pp. 296-302; also *The New York Physician and American Medicine*, May 1956, vol. 46, pp. 42-44; Smith and Bartlett, *Catalogue of drugs and medicines, instruments and utensils, dyestuffs, groceries, and painters' colours, imported, prepared, and sold, by Smith and Bartlett, at their druggists store and apothecaries shop*, Boston, 1795.

⁷⁵ *London Mercury*, London, August 19-26, 1721.

⁷⁶ Bernard Bailyn, *The New England merchants in the seventeenth century*, Cambridge, Massachusetts, 1955, pp. 35-36.

77 *Daily Advertiser*, London, September 23, 1743.

78 "Dr. Bateman's Drops" (see footnote 7).

79 Turlington, *op. cit.* (footnote 15).

80 *Massachusetts Gazette*, Boston, December 21, 1769.

81 John Wesley, *Primitive physic*, 21st ed., London, 1785; *ibid.*, 22nd ed., London, 1788; *ibid.*, 16th Amer. ed., Trenton, 1788; *ibid.*, 22nd Amer. ed., Philadelphia, 1791; George Dock, "The 'primitive physic' of Rev. John Wesley," *Journal of the American Medical Association*, February 20, 1915, vol. 64, pp. 629-638.

82 Rantoul, *op. cit.* (footnote 72).

83 William A. Brewer, "Reminiscences of an old pharmacist." *Pharmaceutical Record*, August 1, 1884, vol. 4, p. 326.

84 *Democratic Press*, Philadelphia, July 1 and October 28, 1824; Thomas W. Dyott, *An exposition of the system of moral and mental abor, established at the glass factory of Dyottsville*, Philadelphia, 1833; and Joseph D. Weeks, "Reports on the manufacture of glass," *Report of the manufactures of the United States at the tenth census*, Washington, D. C, 1883.

85 Van Rensscalar, *op. cit.*, (footnote 53), p. 151.

86 Rantoul, *op. cit.* (footnote 72).

87 *Philadelphia Medical Museum*, new ser., vol. 1, p. 130, 1811.

⁸⁸ *Formulae selectae; or a collection of prescriptions of eminent physicians, and the most celebrated patent medicines*, New York, 1818; John Ayrton Paris, *Pharmacologia; or the history of medicinal substances, with a view to establish the art of prescribing and of composing extemporaneous formulae upon fixed and scientific principles*, New York, 1822.

⁸⁹ Philadelphia College of Pharmacy, *Formulae for the preparation of eight patent medicines, adopted by the Philadelphia College of Pharmacy*, May 4, 1824; Joseph W. England, ed., *The first century of the Philadelphia College of Pharmacy, 1821-1921*, Philadelphia, 1922.

⁹⁰ "Patent medicines," *Journal of the Philadelphia College of Pharmacy*, April 1833, vol. 5, pp. 20-31.

⁹¹ C. Ellis, "Patent medicines," *American Journal of Pharmacy*, April 1839, new ser., vol. 5, pp. 67-74.

⁹² England, *op. cit.* (footnote 89), pp. 73, 103.

⁹³ Carpenter, *op. cit.* (footnote 73).

⁹⁴ William Euen, *A short exposé on quackery ... or, introduction of his son to physicians and country merchants*, Philadelphia, 1840.

⁹⁵ James Winchell Forbes, "The memoirs of an American pharmacist," *Midland Druggist and Pharmaceutical Review*, 1911, vol. 45, pp. 388-395.

⁹⁶ John Uri Lloyd, "Eclectic fads," *Eclectic Medical Journal*, October 1921, vol. 81, p. 2.

[97] Cody & Johnson Drug Co., Apothecary daybooks, Watertown, Wisconsin [1851-1872]. Manuscript originals preserved in the State Historical Society of Wisconsin, cataloged under "Cady."

[98] Swarthout and Silsbee, Druggists daybook, Columbus, Wisconsin [1852-1853]. Manuscript original preserved in the State Historical Society of Wisconsin.

[99] McClaughry and Tyler, Invoice book, Fountain Green, Illinois [1860-1877]. Manuscript original preserved in the Illinois State Historical Society, Springfield.

[100] Harold A. Innis, *Peter Pond, fur trader and adventurer*, Toronto, 1930.

[101] Peter Oliver, "Notes on science, medicine and public health in the United States in the year 1800," *Bulletin of the History of Medicine*. 1944, vol. 16, p. 129.

[102] Isaac Lionberger, "Advertisements in the Missouri Gazette, 1808-1811," *Missouri Historical Society Collections*, 1928-1931, vol. 6, p. 21.

[103] Wedel and Griffenhagen, *op. cit.* (footnote 54).

[104] A. McDonnell, *Contributions to the Historical Society of Montana*, 1941, vol. 10, pp. 202, 217.

[105] *California Daily Courier*, San Francisco, April 25, 1851.

[106] *Political Examiner*, Frederick, Maryland, April 19, 1837.

[107] *Frederick Examiner*, Frederick, Maryland, January 31, 1844.

[108] *Massachusetts Supreme Court*, Thomson vs. Winchester, 19 Pick (Mass.), p. 214, March 1837.

[109] James Harvey Young, "Patent medicines: the early post-frontier phase," *Journal of the Illinois State Historical Society*, Autumn 1953, vol. 46, pp. 254-264.

[110] Cody and Johnson Drug Co., *op. cit.* (footnote 97).

[111] Van Schaack, Stevenson & Reid, *Annual prices current*, Chicago, 1875; Morrison, Plummer & Co., *Price current of drugs, chemicals, oils, glassware, patent medicines, druggists sundries ...*, Chicago, 1880.

[112] Hagerty Bros. & Co., *Catalogue of Druggists' glassware, sundries, fancy goods, etc.*, New York, 1879; Whitall, Tatum & Co., *Annual price list*, Millville, New Jersey, 1898.

[113] Emil Hiss, *Thesaurus of proprietary preparations and pharmaceutical specialties*, Chicago, 1899, p. 12.

[114] Robert B. Nixon, Jr., *Corner druggist*, New York, 1941, p. 68.

[115] Letter from Charles Leich & Co. to Harvey Washington Wiley, Bureau of Chemistry, Department of Agriculture, November 2, 1906. Manuscript original in Record Group 97, National Archives, Washington, D. C.

[116] *American Druggist and Pharmaceutical Record*, 1906, vol. 49, pp. 343-344.

[117] Department of Agriculture, Bureau of Chemistry, Notices of Judgment under the Food and Drugs Act, Notice of Judgment 6222, United States vs. Pabst Pure Extract Co., 1919.

[118] Original handbill, distributed by Standard Drug Co., Elizabeth City, North Carolina, 1925, preserved in the files of the Bureau of Investigation, American Medical Association, Chicago, Ill.

[119] Multiple seizures were made of products shipped by the Horace B. Taylor Co., Fore & Co., and the American Synthetic Co. The quotations are from Notice of Judgment 8868; see also 8881, 8914, 8936, 8956, 8974, 9134, 9147, 9203, 9510, 9586, 9785, 10203, 10204, 10629, 11519, 11669.

[120] Federal Security Agency, Food and Drug Administration, Notice of Judgment 31134, United States vs. McKesson and Robbins, Inc., Murray Division, 1942.

[121] John William De Forest, *Miss Ravenel's conversion from secession to loyalty*, New York, 1867.

[122] Charles H. LaWall, *The curious lore of drugs and medicines (Four thousand years of pharmacy)*, Garden City, New York, 1927, p. 281.

[123] W. B. Sissons, "Poisoning from Godfrey's Cordial," *Journal of the American Medical Association*, March 2, 1912, vol. 58, p. 650.

[124] Edward Kremers and George Urdang, *History of pharmacy*, Philadelphia, 1951, pp. 170, 278.

[125] "Godfrey's Cordial," *Hygeia*, October 1931, vol. 9, p. 1050.

[126] *The dispensatory of the United States of America*, 25th ed., Philadelphia, 1955, p. 158.

[127] *The Pharmaceutical recipe book*, 2nd ed., American Pharmaceutical Association, 1936, p. 121.

[128] Eric W. Martin and E. Fullerton Cook, editors, *Remington's practice of pharmacy*, 11th ed., Easton, Pennsylvania, 1956, p. 286.

[129] Letter from Owen H. Waller, editor of *The Chemist and Druggist*, to George Griffenhagen, January 15, 1957.

www.ingramcontent.com/pod-product-compliance
Lightning Source LLC
Chambersburg PA
CBHW030453220526
45464CB00006B/2512